D1161493

The Hollingworth Letters

This is Number 6 in a series of monographs in the history of technology and culture published jointly by the Society for the History of Technology and The M.I.T. Press.

Publications in the series include:

History of the Lathe to 1850
Robert S. Woodbury

English Land Measuring to 1800: Instruments and Practices
A. W. Richeson

The Development of Technical Education in France 1500-1850
Frederick B. Artz

Sources for the History of the Science of Steel 1532-1786
Cyril Stanley Smith

Bibliography of the History of Technology
Eugene S. Ferguson

The Hollingworth Letters: Technical Change in the Textile Industry, 1826-1837
Thomas W. Leavitt, editor

The Hollingworth Letters:

TECHNICAL CHANGE IN THE TEXTILE INDUSTRY, 1826-1837

Thomas W. Leavitt, editor

Published jointly by
The Society for the History of Technology
and
The M.I.T. Press
Cambridge, Massachusetts, and London, England

Designed by Dwight E. Agner, and
set in IBM Journal by Williams Graphic Service.
Printed by Halliday Lithograph Corp., and bound in the
United States of America by The Colonial Press Inc.

SBN 262 12030 5 (hardcover)

Library of Congress catalog card number: 72-90751

Contents

List of Letters
[vii]

List of Illustrations
[xi]

Acknowledgments
[xiii]

Introduction
[xv]

A Note About Textual Policy
[xxvii]

The Hollingworth Letters
[1]

Genealogical Chart
[110]

Bibliography
[113]

Index
[119]

Letters

From	*To*	*Date*
Jabez Hollingworth Liverpool, England	William Rawcliff	October 9, 1826
John Hollingworth South Leicester, Mass.	William Rawcliff	April 1, 1827
John Hollingworth South Leicester	William Rawcliff	April or May 1827
Jabez Hollingworth South Leicester	William Rawcliff	1827
John P. Barnett Liverpool	William Rawcliff	August 11, 1827
J. Wadsworth Poughkeepsie, N.Y.	William Rawcliff	March 7, 1828
Joseph Hollingworth South Leicester	William Rawcliff	May 20, 1828
George Hollingworth South Leicester	William Rawcliff	June 28, 1828
Joseph Hollingworth South Leicester	William Rawcliff	September 21, 1828
Joseph Hollingworth South Leicester	William Rawcliff	December 7, 1828

Joseph Hollingworth South Leicester	William Rawcliff	February 8, 1829
Joseph and Jabez Hollingworth South Leicester	William Rawcliff	September 6, 1829
George Hollingworth South Leicester	William Rawcliff	October 21, 1829
Joseph Hollingworth South Leicester	William Rawcliff	November 7, 1829
Jabez, George, and Joseph Hollingworth South Leicester	William Rawcliff	January 15 and 17, 1830
George Hollingworth South Leicester	William Rawcliff	January 24, 1830
Joseph and Jabez Hollingworth South Leicester	William Rawcliff	March 13 and 14, 1830
John, Jabez, and James Hollingworth Unknown	Tiffany, Sayles and Hitchcock	April 1830
John, Jabez, and James Hollingworth South Leicester	Tiffany, Sayles and Hitchcock	April 15, 1830
Tiffany, Sayles and Hitchcock Boston, Mass.	John, Jabez, and James Hollingworth	April 16, 1830
Joseph Hollingworth Southbridge, Mass.	William Rawcliff	April 18, 1830
Bradley Clay Huddersfield, England	William Rawcliff	June 5, 1830
John Hollingworth Woodstock, Conn.	William Rawcliff	July 4, 1830

Joseph Haigh Pittsburgh, Pa.	William Rawcliff	July 8, 1830
Joseph Hollingworth Woodstock	William Rawcliff	September 5, 1830
Bradley Clay Huddersfield	William Rawcliff	September 7, 1830
Joseph Hollingworth Southbridge	William Rawcliff	November 2, 1830
George, Jabez, and Joseph Hollingworth Woodstock	William Rawcliff	February 27, 1831
Joseph Hollingworth Woodstock	William Rawcliff	July 17, 1831
Bradley Clay Huddersfield	William Rawcliff	August 1, 1831
John Hollingworth Woodstock	S. A. Hitchcock	October 10, 1831
Joseph Hollingworth Woodstock	William Rawcliff	December 20, 1831
Jabez Hollingworth Sturbridge, Mass.	William Rawcliff	October 7, 1832
John Hollingworth Woodstock	Hamilton Woolen Company	May 2, 1837

Illustrations

Plate I
The Southbridge Light Infantry, on the Common, 1826.
Pen and water color, by C. L. Ammidown. Gift of
Edgar W. and Bernice C. Garbisch. Courtesy of the
Columbus Gallery of Fine Arts, Columbus, Ohio.
[33]

Plate II
Shearing in a Huddersfield shop, ca. 1815.
In the *History of the Huddersfield Woollen Industry*
by W. B. Crump and Gertrude Ghorbal
(Huddersfield, 1935). Courtesy of the Tolson
Memorial Museum, Huddersfield, England.
[41]

Plate III
Dennison Hill, Southbridge, Mass., ca. 1825.
Oil, artist unknown. Courtesy of the National
Gallery of Art, Washington, D. C.
[47]

Plate IV
Globe Village, Southbridge, Mass., 1822.
Oil, by Francis Alexander. Courtesy of the Jacob
Edwards Memorial Library, Southbridge, Mass.
[75]

Acknowledgments

The editor of even a little book needs help from several people. I am particularly indebted to Marion Hyde, who transcribed these letters with me, and to James C. Hippen, who assisted me with the large amount of technical information that these letters required. My thanks also to Betty Goddard, Rex Dennis Parady, Mary Anna Tien, Etta Falkner, and town clerks in Woodstock, Connecticut; Southbridge, Massachusetts; and McDonough, New York. I owe a special debt to the trustees of the Merrimack Valley Textile Museum, who have insisted that the director of the museum should have time to think and to write; consequently I have been able to do so while they raised the money to pay our bills, including a subsidy for this book.

Thomas W. Leavitt

Early in October 1826, a young English laborer took a coach from Huddersfield to Liverpool, where he boarded an American ship bound for New York. The first member of his immediate family to emigrate, Jabez Hollingworth was soon joined by his brothers and sister, mother, father, an aunt and uncle, several cousins, and various acquaintances. Often separated during their migration from place to place, several members of the family stayed in touch with each other by writing letters. Their correspondence survived the Hollingworths and eventually fell into the hands of Alvin Lohr of Hagerstown, Maryland, a bookseller whose offerings are usually the worse for wear. Late in June 1963, he sent the letters to the Merrimack Valley Textile Museum, accompanied by the following note:

> Here are some letters I am sure will be of interest. Sending them along for your inspection as to [*sic*] hard to describe. A fine lot for research. Seems a little high [he was asking $250] but much better the [*sic*] 100 or more common short letters with poor contents. These tell complete history like a book around 135 years ago & are well written and easy to read.

Lohr was telling the truth. As soon as he had read the letters, Rex Dennis Parady, the museum librarian, decided the museum should buy them. The decision was a sound one, for the letters offer a rare view of the immigrant labor force in Jacksonian America. Moreover, they are a singularly appropriate acquisition for a museum whose purpose is to study the role of textile manufacturing in American history.

At first glance, there appears to be little that is unique about this particular family of immigrants. They were, after all, part of a tradition as old as the colonization of North America. Edward Johnson claimed that the settlers of Rowley, Massachusetts, had in 1638 been

> . . . the first people that set upon making cloth in this western world; for which end they built a fulling-mill, and caused their little ones to be very diligent in spinning cotton-woole, many of them having been clothiers in England till their zeale to promote the gospel of Christ caused them to wander.[1]

Rowley's settlers were only a few among thousands of Englishmen with textile skills who emigrated to the North Atlantic colonies before the factory system assumed its central place in American history. John C. Hotten's *Original Lists of Emigrants, 1600-1700,* suggests that nearly every passenger ship carried a few professional weavers and dyers aboard.[2] And thanks to the painstaking research of William R. Bagnall,[3] we know of many others who brought clothmaking skills with them to America in the Colonial and early Federal periods. The Scholfield brothers of Yorkshire and Samuel Slater of Derbyshire are probably the most famous of these earlier immigrants.[4] The contributions of these immigrant families are deservedly well known, for they played unique roles in what has been labeled "the transit of technology."[5] A different fate awaited the Hollingworths, who

---

[1]*Wonder-Working Providences of Sion's Saviour in New England* (London, 1654), pp. 131-132, as quoted in William R. Bagnall, *The Textile Industries of the United States* (Cambridge, Mass., 1893), 1: 2.

[2](London, 1874), passim.

[3]Bagnall, *Textile Industries,* passim.

[4]*Ibid.*, pp. 135 ff., 202 ff.

[5]Carroll W. Pursell, Jr., "Thomas Digges and William Pearce: An Example of the Transit of Technology," *William and Mary Quarterly,* 3d ser. 21, no. 4 (October 1964): 551-560.

arrived when the factory was already a common feature of the American scene. Like hundreds of their contemporaries the Hollingworths joined the system and used it as a means of improving their economic and social standing. It is because the Hollingworths are, statistically at least, so common, that their letters are valuable to us. They were among the first generation of an "accelerated emigration" to America that began about 1815, causing an "appreciable drain" on the population of England, according to Arthur W. Redford.[6]

The Hollingworths left the West Riding of Yorkshire during the depression which began in 1825 and from which England did not recover until after 1833. Unfortunately, we cannot know all of their immediate motives for migrating. The depression was undoubtedly a factor, but their letters suggest another, more positive reason for wishing to change their environment: the desire to live in America where—as others reported—life was more abundantly pleasant.[7] For the area in and around Huddersfield, where the Hollingworths lived (and where, for generations, their ancestors had lived), was being transformed by a revolution in wool technology. Furthermore, they wanted to get away from the "tyrannical" factory system that was part of England's unreformed political and social order.

---

[6]Arthur W. Redford, *Labour Migration in England, 1800-1850,* 2d ed., edited by W. H. Chaloner (Manchester: Manchester University Press, 1964), pp. 2, 173. Actually, the tide ebbed and flowed: the annual number leaving Great Britain in the ten-year period beginning in 1825 ranged as high as 5,352 in 1828 to a low of 1,153 in 1830 and then to a new high of 10,490 in 1834. U.S. Bureau of the Census, *Historical Statistics of the United States. Colonial Times to 1957* (Washington, D. C., 1960), p. 57.

[7]G. Poulett Scrope, *Extracts of Letters From Poor Persons Who Emigrated Last Year to Canada and the United States,* 2d ed. (London, 1832), passim.

In the West Riding of Yorkshire fully two-thirds of the population depended on the wool trade.[8] The amount of cloth produced there had increased substantially in the fifty years before 1820. But the number of people employed had not grown proportionately; mechanical improvements had made it possible for one man to do more work. Before 1787 carding by hand had lost its former significance: water wheels turned the carding machines. By 1800 the spinning jenny was in general use, and slubbing was done by the billy. The power mule, although new, bore in its frame the threat of a fully mechanized spinning process. Almost all weaving was still done by hand, but power-driven machinery was radically changing the finishing processes. It was the revolution in the techniques of cloth finishing that most disturbed the equanimity of the countryside

---

[8] Abraham Rees, "Woollen Manufacture," *The Cyclopaedia; or Universal Dictionary of Arts, Sciences and Literature*, vol. 40 (Philadelphia, 1822). W. B. Crump and Gertrude Ghorbal, *History of the Huddersfield Woollen Industry* (Huddersfield: Tolson Memorial Museum, 1935), pp. 64-70, 118-119. The principal steps in the manufacture of woolen cloth were common knowledge among the Hollingworths and should be kept in mind while reading their letters:

1. Scouring and picking: washing and untangling the fibers.
2. Carding: further untangling, cleaning, and partial parallelization of the fibers.
3. Condensing: preparation of the fibers for spinning by formation into a loose strand called roving or roping.
4. Spinning: twisting and attenuation of the roving to form yarn.
5. Weaving: interlacing two sets of yarn—warp and weft—to form cloth.
6. Finishing: a general term applied to fulling, napping, shearing, pressing, and dyeing the cloth after it is woven.

I have postponed further explanations of these processes and of the machines associated with them until the appropriate place in the letters which follow. See these footnotes: 5, 12, 20, 23, 24, 25, 26, 35, 36, 41, 49, 50, 54, 55, 66.

around Huddersfield. In 1812 the croppers rioted when the shearing frame threatened to supersede the hand shears they had traditionally used to finish cloth. The change in technology would of itself have disturbed the finishers, but in this instance it coincided with rising food prices, falling wages, and increased unemployment, caused by the naval blockades that were part of the Napoleonic Wars. The opposition to the introduction of machines in the finishing processes signaled the imminent defeat of the domestic system as it gave way before the factories. The independent artisan and small manufacturer were on their way out.[9] After 1800 periods of prosperity were followed regularly by periods of depression. It was the recession of 1826 that provided the Hollingworths with the excuse they needed to pull up their English roots and replant them on the other side of the Atlantic.

In America the major processes in wool manufacturing had been mechanized: when the Age of Jackson dawned, the first stage of the Industrial Revolution was complete. Already Americans were exporting improved techniques of clothmaking to Great Britain, while at home, tiny farming communities were being transformed almost overnight into thriving factory

---

[9]The pace of the revolution was uneven and in parts very slow. As late as 1832 an American tourist in Huddersfield could say, "Here I found the great and ponderous shear-blades in common use, where the light circular revolving shear-blade would have performed the same work in a more perfect manner, at half the expense for attendance. I also witnessed in an old building several men employed in raising the nap on broadcloth by teasles, applied by manual labor, instead of being attached in the now common way to the surface of the revolving barrel of a cylinder or drum, called a gig-mill, whereby two men are enabled to perform more work than a dozen in the old manner." Zachariah Allen, *The Practical Tourist, or Sketches of the State of the Useful Arts, and of Society, Scenery, &c. &c. in Great-Britain, France and Holland* (Providence, R.I., 1832), 1: 190.

villages. One contemporary estimate was that 60,000 people were engaged in wool manufacturing.[10] An English visitor remarked

> [The Americans] . . . seem determined to have a prosperity of their own making; to set up rival Birminghams and Manchesters; and, in spite of "nature and their stars," to become, without delay, a great manufacturing, as well as a great agricultural nation.[11]

The various members of the Hollingworth family were well equipped to help promote manufacturing in America. Almost instinctively they gravitated toward central New England, where the machine shop and factory were already part of everyday life. Consequently, their letters tell us a great deal about the effects that the new technology had on labor. In this dynamic environment there was no such thing as job security. The only kind of security possessed by the individual laborer was the knowledge that if he lost one job he might find another. He could also be confident that a particular skill might be useful outside as well as inside the factory. And, if no factory work was available, he could turn to the land.

---

[10] *Niles' Weekly Register,* April 21, 1827, p. 139. The editor boasted, "We have more than once noticed exportations of machinery to Great Britain. The power loom made at Matteawan, N.Y. for weaving broad cloth, and sent out some time ago, has been put up and is in operation at Leeds, highly approved of. In the knowledge of machinery, we are fully able to return 'light for light' " (November 24, 1827, p. 195). Earlier the same year, a resident of Orange County, N.Y., reported in a letter to the editor that the village of Walden, which consisted of only three farms in 1822, presently had three textile mills, which processed 100,000 lbs. of wool annually (May 19, 1827, p. 194).

[11] [Thomas Hamilton,] *Men and Manners in America* (Philadelphia, 1833), p. 109.

Jabez, for example, appears at first glance to be a machinist, but on closer examination one discovers he is primarily a carpenter. Even as he finds work inside a machine shop, his family is discussing the desirability of sending him to New York to build houses. In order to prepare for this eventuality, he uses part of his earnings to buy the tools he would need for this or a similar job. Apparently a competent laborer, Jabez is able, within a year, to command a $5 raise from $20 to $25 a month. Yet his usefulness is limited, for before another year is up, Jabez finds himself replaced, as his father reports, by a man "that can work boath in Wood and Iron. This is rarely to be found except in a Yankee who professes to do every thing. . . ." Unemployed because iron was replacing wood in textile machinery, Jabez turns to making wagons and a hand loom for domestic use. When we last catch a glimpse of him, Jabez has moved again, this time to Sturbridge, where he is erecting a meetinghouse for Baptists. Meanwhile, when John Hollingworth loses his job in the mill, he takes up farming.[12]

Every laborer was subject to the arbitrary will of the mill owner or the superintendent. If machinery needed to be repaired on Sunday—when the mill was otherwise generally unoccupied—the machinist went to work. If the boss spinner wanted yarn made on Sunday evening, the laborer was faced with little choice but to do it. If the shearers slipped behind in production, they worked "night and day" until they caught up. When orders were plentiful, the entire mill force worked from moonlight to moonlight. And when the mill owner discovered that girls could do jobs traditionally held by adult males, the

[12]Letters of April 1, 1827; [undated] 1827; June 28, 1828; September 6, 1829; October 21, 1829; January 17, 1830; March 14, 1830; October 7, 1832.

men were fired, "quality being out of the question," as Jabez remarks.[13]

The same environment that so often worked to the disadvantage of the individual laborer was filled with opportunity for the entrepreneur. The same man often played both roles. Here, too, the experience of the Hollingworths is typical of what happened to a great many people. Whether they are temporarily well situated, unemployed, or en route to new jobs, the Hollingworths keep always in front of them the vision of their own factory, which would enable them not to escape the system but to be victorious over it. They are mill managers in spirit long before the fact. At the beginning of their American sojourn, several members of the family talk about organizing a communal society modeled on Robert Owen's community at New Lanark, Scotland, and at New Harmony, Indiana. They locate a suitable farm in western New York, but as they discuss the possibility of moving onto it, they keep looking for available factories. Although credit is obtainable, they are reluctant to assume too large a risk with the relatively small amount of capital that they possess. They explore the possibility of strengthening their position by bringing in other families. Finally, they settle on a factory in Woodstock, Connecticut, which they lease for a three-year term. Meanwhile their cousins and acquaintances look for similar situations, which would allow them to change their roles from employee to employer. They are "on the make" from the day when they land on American soil.[14]

---

[13]Letters of May 20, 1828; September 21, 1828; March 14, 1830.

[14]Letters of April 1, 1827; [undated] 1827; October 21, 1829; January 15, 1830; January 17, 1830; January 24, 1830; March 14, 1830; April 15, 1830; July 8, 1830; November 2, 1830.

Tocqueville commented, ". . . democracy not only swells the number

What also seeps through this record is the critical role of the Boston firms that supplied capital and commercial connections for these mills. The reader may almost feel the all-embracing hand from Boston's financial district reaching out to furnish the mills with raw material and distribute their finished products. Sensitive to the fortunes of these mills, the men who run the commission houses often find it makes sense to buy out a hard-pressed owner and then to lease the property to hungry entrepreneurs like the Hollingworths. In this way the Boston men accumulate the money and experience that will make it possible for some of them to become rich.[15]

The American communities in which the Hollingworths lived were, without exception, places with just a few hundred families. Their common characteristic, in addition to their small size, was the fact that they encompassed streams of water large enough and dependable enough to turn the water wheels and turbines that drove the mill machinery. It was in this milieu that the Hollingworths sought to establish themselves. Their experience, in fact, suggests that New England's mill towns were—perhaps to a greater degree than we have realized—dynamic commercial and political units.

The family settled initially in Leicester, Massachusetts, a town with a population of about 1,200 people. John and Jabez, the first to arrive, went to work for the Leicester Manufacturing

---

of working men, but it leads men to prefer one kind of labour to another; and while it diverts them from agriculture, it encourages their taste for commerce and manufactures." *Democracy in America,* trans. Henry Reeve, preface and notes by John C. Spencer, 4th ed. (New York, 1841), 2:164.

[15]Letters of April 1, 1827; August 11, 1827; October 21, 1829; March 14, 1830; April 15, 1830; April 16, 1830; September 5, 1830; October 10, 1831; May 2, 1837.

Company, which made woolen cloth. The mill complex included 3 major buildings, the largest of which was 4 stories high, 100 feet long, and 40 feet wide. The company also owned a dye house, a store, and 11 homes for employees.[16] In December 1827, Joseph, James, Edwin, and their father, George, joined John and Jabez in Leicester. When they arrived, they found Jabez in the machine shop and John slubbing. Joseph went to work in the finishing room, James spun, and Edwin, the youngest, served as a warp winder. Mr. Hollingworth apparently worked as a weaver.

Spring of 1830 found Joseph, James, Edwin, and George Hollingworth in Southbridge, Massachusetts, a town with about 1,100 people, where they were employed by the Hamilton Woolen Company, then one of the largest firms in America. Father and Joseph were warping, James was a jack spinner, and Edwin was spooling. The company, which produced broadcloth, employed more than 100 people. The mill contained 5 sets of cards, dyeing and finishing apparatus, and 28 broad looms capable of manufacturing 40,000 yards of cloth annually.[17]

John and Jabez moved to Woodstock, Connecticut (population ca. 2,900), about the same time to work in the Muddy Brook-Pond Factory, which they leased.[18] They were joined within a year by their other brothers and their father. Their

---

[16]The material for this and the succeeding paragraphs about Leicester is taken from Jeremiah Spofford, *A Gazetteer of Massachusetts* (Newburyport, 1828), p. 226; and *History of Worcester County, Massachusetts . . .* (Boston, 1879), 1: 631-632.

[17]"Brief Record of the History of the Hamilton Woolen Company, Southbridge, Massachusetts," filed with the company records, Baker Library, Harvard University; Spofford, *Gazetteer of Massachusetts*, p. 297; Holmes Ammidown, *Historical Collections* (New York, 1874), 2: 374; *History of Worcester County* 2: 295-313.

[18]Clarence Winthrop Bowen, *The History of Woodstock, Connecticut* (Norwood, Mass.: privately printed, 1926), pp. 226, 545.

attempt to run the Woodstock factory as a family venture is the last reasonably full account we have of their life in America. The family apparently kept the factory for the original three-year lease, but they did not purchase the mill. Jabez moved to Sturbridge sometime during 1832 and, with his father, disappeared from the record. James Hollingworth moved back to Southbridge, where his children were born over the course of the next decade. Edwin temporarily submerged, but appeared again in Waterford, Connecticut, as a young father before moving to McDonough, New York, where as late as 1874, E. Hollingworth and Son were manufacturing cassimeres and flannels. John stayed in Woodstock, where he bought a farm in 1835. He was also apparently involved in the bankruptcy proceedings of the Muddy Brook Manufacturing Company in 1837 when he sold some machinery to the Hamilton Woolen Company, according to the final note in this collection. Joseph also remained in Woodstock, where at various times he owned a blacksmith shop, another woolen factory, and a sawmill. He died there on October 9, 1861.[19]

Thus the fate of the Hollingworth family in America is to disperse gradually and to find diverse occupations. Opportunity, for them, appears under the guise of change and innovation. Having left Yorkshire in order to escape the English factory system, they try to turn the American factory system to their advantage. They are not wholly successful, but in their attempt to do so they have left us a record we can profitably study.

---

[19]The vitae on James, Edwin, John, and Joseph are from the following sources: Southbridge, Mass., *Vital Records;* McDonough, New York, *Vital Records; United States Textile Manufacturers' Directory* (Boston, 1874), p. 185; Woodstock, Conn., *Vital Records;* Woodstock, Conn., *Land Records* 21: 147; 23: 259, 325; 24: 316, 337-338, 407; J. Leander Bishop, *A History of American Manufactures from 1608 to 1860* (Philadelphia, 1864), 2: 794; Bowen, *History of Woodstock*, p. 226.

A Note About Textual Policy

I have generally followed the rules that L. H. Butterfield, editor-in-chief of The Adams Papers, outlined on pp. lv-lix of his Introduction to the *Diary and Autobiography of John Adams* (Cambridge: Harvard University Press, 1961). The rules are summarized here:

1. Original spelling is preserved as found in the manuscript, but slips of the pen are silently corrected.

2. Grammar and syntax are preserved as found in the manuscript.

3. Capitalization is preserved as found in the manuscript except that (1) all sentences begin with capital letters, (2) all personal and geographical names are capitalized, and (3) where the author's intention is not clear, modern usage is followed.

4. Punctuation is normally preserved as found in the manuscript, but a few rules of conventionalization have been applied: (1) every sentence ends with a period, (2) dashes intended to be terminal marks are converted to periods, superfluous dashes are removed, and (3) intrusive commas are omitted.

5. Abbreviations and contractions are preserved as found if they are still in use or are readily recognizable by a modern reader.

6. Missing and illegible matter is indicated by square brackets enclosing the editor's conjectural readings or by suspension points if no reading can be given. If only a portion of a word is missing, it may be silently supplied when there is no doubt about the reading.

The Hollingworth Letters

Jabez Hollingworth to William Rawcliff

Liverpool Octr 9th 1826

Dear Uncle

This is to inform that I have taken my passage in the American Ship Hamilton, burden 450 Tons, Captain G Bunker the Master, bound for New York.[1] I told you in my last letter that I should let you know how I laid out my money. I shall now give you a clear statement.

	£		S		D
Coach Fare from Huddersfield to Manchester			5	..	0
Box from ditto			4	..	0
Lodging and eating in Manchester			3	..	0
Coach Fare from Manchester to Liverpool			4	..	0
Box ditto			4	..	0
Passage Money	5	..	5	..	0
Entrance at the Custom House		..	2	..	0
Bread and bag		..	8	..	0
Potatoes and bag		..	6	..	11
13¼ lbs of Ham at 9d per lb.		..	9	..	11
33 lbs of oat Meal		..	6	..	0
7½ lbs Sugar at 7½d per lb.		..	2	..	6
1 lb Currants and 1 lb Raisins		..	1	..	10
	8	..	3	..	8

[1]The Hamilton saw temporary service with the New York to Liverpool New Line or Red Star Line, whose New York agents were Byrnes, Trumble & Company. The Liverpool agents were Cropper, Benson & Co. Carl C. Cutler, *Queens of the Western Ocean* (Annapolis, Md.: U.S. Naval Institute, 1961), p. 378.

	£		S		D
Brought over	8	..	3	..	8
1 lb Soap		..		..	7½
½ lb Coffe and 2 oz of Tea		..	2	..	4
½ lb of Epsom Salts		..		..	8
Apples and Onions		..	1	..	6½
Meal bag and net		..	1	..	2
Tins Pots &c.		..	3	..	1
Razor Box and Brush		..	4	..	0
Hairy Cap		..	1	..	6
Straw Bed		..	1	..	0
4th Cheese at 8d per lb		..	2	..	8
1 lb Candles		..		..	7
2 lb Butter		..	2	..	0
8 lb Treacle at 3½ per lb		..	2	..	4
1 quart Brandy		..	9	..	6
Carriage of Luggage to the Ship		..	1	..	9
Lodging and Eating at Mr Witakers		..	11	..	0
Addition of Bread		..	2	..	4
Pewter Plate		..	1	..	3
What I cannot give account for about		..	5	..	0
	£10	..	18	..	0

I have now given you as clear a statement as possible. I have about 2 £ 5 d. left and I mean to get them exchanged for American Dollars. I am in good health and I should be in good spirits if my relations was with me, but I feel it a very hard task to leave them behind me. We expect to sail to morrow if the wind be fair. You must excuse me for writing so little for I am so bothered that I can think of nothing. I shall sleep on board the first time to night.

Give my respects to my Father and Mother and Brothers and to brother Johns wife and to all enquiring Friends.

I am with respect
Your ever loving Nephew
Jabez Hollingworth

[To] Mr William Rawcliff
Oldfield near
Huddersfield
Yorkshire

⁂ ⁂ ⁂

John Hollingworth to William Rawcliff

South Leicester April 1st 1827

Dear Uncle

I take the opportunity of writing to you to let you know that my Brother Jabez and I are in a good state of health at present and that we are doing all that lays in our power to accomplish the objects that induced us to leave Old England to brave the dangers of the Atlantic Ocean and to come to America that is to provide an happy asylum for our kindred our friends and ourselves, and it is our firm opinion that it will be best to form ourselves into a system after the manner of Robert Owen's plan,[2] that is to form ourselves into a society in com-

[2]Robert Owen (1771-1858), after initiating reforms at his cotton mills in New Lanark, Scotland, had come to America in 1825 with his son, Robert Dale (1801-1877), to establish a communal society at New Harmony, Indiana. At the moment when John Hollingworth was writing this letter, the Owens were about to conclude that the experiment at New Harmony was a failure. *Dictionary of National Biography*, vol. 14; *Dictionary of American Biography*, vol. 7.

mon to help and assist each other and to have one common stock for it is the unatural ideas of thine and mine that produces all the evils of tyranny slavery poverty and oppression of the present day. My Brother Jabez, Cousins George and James Hollingworth Joseph Kenyon and I have made a contract for a parcel of land consisting of 200 Acres together with a house and other nessasaryes and there is 24 Acres of it cleared. All situated in the Township of Yorkshire in the County of Chataraugus in the State of New York within 40 Miles of the Town of Buffalo which is situated on the head of the great Westren Canal. The farm has two public roads crossing each other at right Angles and runing on the east and north sides of the farm. There is also a water privelege on the same and with your assistance we think of erecting a small Factory at some futer period. There is also a branch of the Westren Canal in agitation to come within 2½ miles of the same place.[3] Now we should wish you, my Father, my Uncle John Hollingworth and my Uncle John Kenyon to meet together and discus the subject form such plans as you think most proper and assist each other to utmost of your power for

Take this maxim old and young
Friendship and union makes us strong.

I should earnestly wish you to acceed to our plans and to fall in cooperative with us. I also earnestly wish that Uriah Haigh and James Bates would come and heartley join us in a cause the effect of which would be real happiness to ourselves and our Children after us. My Cousins George and James Hollingworth

[3]This proposed canal was never constructed; instead the Genesee Valley Canal was started in 1836, eventually connecting the Erie Canal at Rochester with the Allegheny River at Olean. Balthasar H. Meyer, *History of Transportation in the United States Before 1860* (Washington, 1898), pp. 202, 654.

has just recived a letter dated Nov 24th 1826 which sayes that my Uncle John Hollingworth much approoves of my plan of cooperation and it is my sincere wish that it may meet the approbation of you all which is what it deserves. It is our intention to each of us to do all that lays in our power to accomplish our object and to keep regular accounts of our expences and when we get setteled to make a fair calculation and see what one has expended more than another so that they may have Intrest according to the money expended and the remainder to be divided equally amongest the members of the Community yearly which they may put in again as stock.[4] As to myself I still am working in the Card Room at Slubing and if all be well I intend to remaine at the same work till we have accomplished our undertaking and by that means I may be usefull at some future time in the same business.[5] My Brother Jabez is still working in the Machine Shop[6] but I expect we shall send him together with another of our company who may be geting the least wages to the farm to be prepareing for the reception of us all which I hope will not be long before we be there and then I

---

[4]These plans for a farm south of the Erie Canal did not come to fruition. See Joseph Hollingworth to William Rawcliff, December 7, 1828.

[5]The carding machine or "card" (hence "card room") consisted of one large and several small water-powered cylinders clothed with thousands of short, protruding wires that separated and straightened the fibers. The process produced short rolls of wool as long as the card was wide. These rolls were prepared for spinning by piecing them together and attenuating them on a hand-powered slubbing billy. While a boy pieced the rolls together at the feed end of the billy, John Hollingworth would manipulate the carriage of the billy to twist and draw out the rolls into a continuous roving.

[6]I.e., in the machine shop that was part of the mill where John was employed.

hope the following lines from Doctor Watts will be applicable to us[7]

Blest are the sons of peace,
Whose hearts and hopes are one,
Whose kind designs to serve and please
Thro' all their actions run.

and again

How pleasant 'tis to see
Kindred and friends agree,
Each in their proper station move,
And each fulfil their part
With sympathizing heart,
In all the cares of life and love!

Give my tender love to my Wife and tell her that although she may yet be in England through a peice of Villany that has been practiced upon me that I have not forgot her, and that she must make herself as content as she possiabley can as I hope we shall not be long before we see one another again in a place where we shall be free from the oppression of the Manufactureing System.

I remaine with respect your most
Affectionate Nephew John Hollingworth

---

[7]Like thousands of other Dissenters' children in the nineteenth century, the Hollingworths were raised on Isaac Watts (1674-1748), Congregational minister, hymn writer, and poet. John Hollingworth apparently brought his copy of Watts's *Psalms, Hymns, and Spiritual Songs* with him, including these verses, which are excerpts from S. M. Peckham's "Union and Peace" and P. M. Dalston's "The Blessings of Friendship." These two hymns are printed in Samuel M. Worcester's New Edition (Boston, 1852), p. 260.

Send me word how my little Mary and William are and how you all are.

N.B. If you should approve of our plan I would advise you to loose no time in disposeing of your property and to come to this country and to do all that layes your power to assist the persons mentioned in this letter likewise. Write imeadeately and let me know your mind on the recipt of this.

[To] Mr William Rawcliffe
Oldfield near
Honly
Yorkshire

❧ ❧ ❧

John Hollingworth to William Rawcliff

P.S. Although my last letter was dated April 1st it was wrote a few days before on account of having to send it by a person at our place who was going to England and having others letters to write I had finished yours and sealed it up when my family arrived at this place in good health. Consequently I had not the chance of informing you of their arrival and of expressing my gratitude to you for your trouble and your assistance to me and my family which I intend to repay as soon as I can conveniently. You say you cannot see what objections I can have for my Father and his family to come to America. I have none only that I should like to have them independant of the factory syestem which cannot be done all at once, but I would rather that they were in the factorys of America then they should be starving in England but I should like you to acquiesce in our plans.

May 3rd Since writing above My Cousin Joseph Kenyon has recieved a letter from his Brother Eli partly wrote by himself and partly by my Father and Uncle John Hollingworth which has put us to a plung as they want to come immeadtly and they want our assistance which we cannot both give and pay for our land but we shall render them all the assistance we can whether we have the land or not but we think we could do both with your assistance. With respect to the land we shall let that remaine as it is untill we receive your answer to our letters dated April 1st and then we shall act according to your advice, but I cannot draft this subject without giving you a few extracts from a report of the committee on inland Navigation and internal improvement, to the Pennsylvania Legislature.[8] The Erie Canal has connected the river Hudson with Lake Erie and in a short time the Ohio Canal will connect the same waters with the extensive vale of the Mississippi and Ohio. One hundred and eighty-three miles of the Ohio Canal will be finished early in the summer of next year, which will be an extension of the Erie Canal through the heart of the state of Ohio.[9] The committee will ask the attention of the House to the interesting fact, that a few years ago, the trade of Lake Erie required but 2 or 3 small sloops, and that during the last year, (the first year after the completion of the Erie Canal,) 6 steam boats and nearly an 100

---

[8]John is referring to the "Report of the Committee on Inland Navigation and Internal Improvement relative to the further extension of the Pennsylvania canal, accompanied with a bill," read in the Pennsylvania House of Representatives on February 28 (Harrisburg, 1827).

[9]The Ohio and Erie Canal left Lake Erie at Cleveland and ran south 309 miles to Portsmouth. There it joined the Ohio River, by which one could reach the Mississippi. Started in 1825, it was not completed until 1832. J. Calvin Smith, *The Western Tourist and Emigrant's Guide* (New York, 1840), p. 72; H. S. Tanner, *A Brief Description of the Canals and Railroads of the United States* (Philadelphia, 1834), p. 59.

coasting vessels were employed upon the lake by the trade which exists between New York and different points of that inland sea. This fact enables us to form a satisfactory judgment as to result of our situation, if we determine to remain stationary. Besides I can give you another proof of the growing and improveing state of the Westren Country occasioned by its New inland Navigation. Last Spring soon after I got to Pleasant Vally [N.Y.][10] when the North River broke up there were only 12 Steame boats which was the usal number and before I left that place which was only 16 weeks they amounted to 27 and the Sloops increased in the same ratio which now amounts to betwixt 2 and 3000. The town of Rochester which I mentioned as being situated on the canal is growing so fast that is said there are great symptoms of it rivaling New York,[11] and the avidity with which every thing in the way of business is carried on in it that all the cellars under the new maket which they have built were all applied for before it was built or even begun of. It has grown so large that the citizens cannot find each other without a Directory which will prove that the westren Country is not altogether a wilderness as you suppose. We know that it is only a few years since it was a wilderness but its natural advantages together with its artifical improovements will not only render it as peopolous but both more peopolous and more desireable to live in than any part of the New England State. Besides I must be plain, the factorys are not to be depended upon. We might

---

[10]Home of a woolen mill owned by the Pleasant Valley Manufacturing Company. Edwin Williams, *The New York Annual Register for the Year of Our Lord 1830* . . . (New York, 1830), p. 149.

[11]Rochester did not, in fact, quite catch up with New York City in this or any later decade. However, due largely to the advent of the Erie Canal, the population did increase by almost 900 per cent between 1820 and 1830. Henry O'Reilly, *Sketches of Rochester* (Rochester, 1838), p. 34.

send for our Parents and their families and not only have a favourable prospect but a full promise of employ for them and when they get here there would be neither house nor work for them which would be great disapoinment to them. But all these evils might be provided against by you joining us that we might both get them to this country and keep the land which would be a home for them provided they did not get into employment when they got here.
Now the most proper plan that could be taken would be for them to sell what they have and what they are short for you to make up if you have it in your power and for you all to come together and by that means you might come for less money than if you were to come seperate and we would get work for as many as we could and the remainder to go on to the farm along with you and my brother Jabez to build a few houses and other building that would be nessarary while we were working to furnish you with money. For we yet remaine assured that no plan but coopperation will secure us Prosperity and happiness either in this or any other Country. You may ask what securety you can have provided you should act according to our proposals. We answer that we will give you the possesion of the land if we hold our bargin (for your money and intrest). If not we shall be nearly able to pay you when you get here, besides there is 3 more besides myself here who conduct themselves in a steady, sober, and Industrious manner whom I should not hesetate one moment to trust with as many thousands Dollars if I had it in my power. Since my last letter I have got into a house to myself and my Brother and Cousins intend to come to me as soon as posiable. My Brother is still working in the mechine Shop where I hope he is learning something that will be usefull to us at some futer time. He says that he shall send you another letter soon. My Cousin James is still weaving at this

place and my Cousin Joseph is Spining.[12] George is at a place 24 miles from here belonging to the same Company.[13] I am still slubing. We are all in good health at present except Son William who is about some teeth but I hope he will soon be better. James Haddock is still at this place but he is a curious kind of character at best. He is only mischief making tell-tale, one to get nobodys good word. I said in my fathers letter that Geo Brown appeared different to what he used to do and that I did not know what to think about him but we have had an explaination and it was all through a parcel of lies of J. Haddock telling who has striven at nothing but making mischife ever since he came to the place but no more upon this subject. We wish you to send an answer to this letter as speedy as possibale. Direct your letters to the Care of Cropper Benson & Co that they may come

---

[12]Cousin Joseph could have been spinning on either a jenny or a jack. The jenny, patented by James Hargreaves in England in 1770, and introduced into America as early as 1775, was a hand-powered machine that attenuated and twisted roving into yarn. Both the jenny and the larger, partially hand-powered, and partially automatic jack spun on the same principle as the common wool spinning wheel ("great wheel"). The jenny was, in effect, some twenty to fifty spinning wheels operated by one man; the jack could have two hundred or more spindles. C. Aspin and S. D. Chapman, *James Hargreaves and the Spinning Jenny* (Helmshore, Lancs.: Helmshore Local Historical Society, 1964), chap. 4; Arthur H. Cole, *The American Wool Manufacture* (Cambridge, Mass.: Harvard University Press, 1926), 1: 108-117. Cousin James may have been working at a hand loom or operating a power loom. Like the spinning process, weaving was in a transitional phase between 1820 and 1830; in the absence of unambiguous evidence, one cannot be certain when a particular factory wholly abandoned hand for machine power.

[13]He is referring to the Saxon Factory in Framingham, Massachusetts. U.S., Congress, House, Committee on Manufactures, *Report on Petition relative to . . . Duties on Imports*, 20th Cong., 1st sess., January 31, 1828, H. Rept. 115, vol. 2, p. 120.

by the New York Packet as we shall get them about a Month Sooner

I remaine with respect your affectionate Nephew
John Hollingworth

[To] M[r] William Rawcliff
Oldfeild near Honley
near Huddersfeild
Yorkshire
England

❀ ❀ ❀

Jabez Hollingworth to William Rawcliff

South Leicester [Mass., 1827]

Dear Uncle & Aunt.

I take this opportunity of writing a few lines to you in good health as I hope they will find you the same. In this letter I shall give you such information concerning America as I am able. When I left England you wished me to look out for a situation for you; as for looking out I never had much opportunity as it has been the winter season, but I have made as much inquiry as I possibly could and I am truly satisfied that the State of New York is and will be the finest State in North America. As Agriculture is the source of human life, I shall treat upon it the first. As to this state it is not very fit for Agriculture as it is chiefly barren, stony, and of a sandy soil. This place is situated near the highest lands in the New England states. Many is going to leave this neighbourhood for the Western country to purchase land and settle. The Western part of the state of New York is a very fine healthy country and very good land. It will produce from 30 to 40 bushels of wheat per acre, which is worth 50 cents per bushel. The taxes of all descriptions will amount to about 8 dollars for every 100 acres of cultivated land, and about 4 dollars for every 100 acres of wood land. I have not yet seen nor heard of any uncultivated land but what grows wood. Land is from 3 to 10 dollars per acre in the Western part of the state of New York according to the quality and situation. As to manufacturing it increases rapidly. I have seen better cloth made here than ever I saw in England. This state is better calculated for manufacturing than farming. This causes it to be more like England, because where manufacturing flourishes Tyrany, Oppression, and Slavery will follow. In order to prove this I will tell you what has past since I came here. Wages has been reduced 25 percent, that is if a man could earn 1 dollar per day, he can now only earn 75 cents per day, but this is better than 10d per day on the roads in England. As to my own part, I am very thankfull that I have escaped that irrisistable misery which England is now in, and that I have got to a country which with Industry Frugality and Perseverance may prove not only to me but to you a happy Asylum. As to the manner of living there is no King on earth can live better. We have everything to eat that a reasonable man can wish for. We have beef or pork 3 times a day potatoes cheese butter and all kinds of bread except oat bread tea and coffee and sometimes milk, and to our sunday dinners we have Indian meal porridge and milk to them. Cyder is our chief drink. I shall now give you a short sketch of the price of provisions both in this state and in the state of New York.

	$	cts Massachusetts	$	cts New York
Flour per barrel	8 ..	0	5 ..	0
Cyder do	1 ..	0	1 ..	0
Beef per lb.	0 ..	6	0 ..	3

Mutton do	0 .. 7	0 .. 4
Veal ditto	0 .. 4	0 .. 1
Pork ditto	0 .. 6	0 .. 3
Butter do	0 .. 16	0 .. 16
Cheese do	0 .. 8	0 .. 6
Sugar do	0 .. 9	0 .. 6
Currants do	0 .. 22	0 .. 22
Brandy per gal.	1 .. 90	1 .. 50
Jamacia Rum	1 .. 20	
West India Rum	1 .. 0	
New England Rum	0 .. 40	

Boots from 3 to 5 dollars per pair
Shoes fr 2 do.
Hats from 3 to 8 dollars a piese

My Brother John, my Cousin George and James, and Joseph Kenyon and I have made an agreement for 200 acres of land in the town ship of Yorkshire in the county of Cattaraugus in the state of New York. It is situated 40 miles S. E. of Buffalo a town situated at the head of the great western canal. It is 90 miles S. West of Rochester a large and flourishing town on the canal. It is 270 miles S. W. of Uitica a very fine and large town on the canal and from its central situation is likely to become the seat of government of the state. The farm contains a house and 24 acres of cleared land, (it has the first crop on at present) and a small water privelege on it, also there is a well sunk, and an excellent top spring. We are to give 700 $ for the said land, we paid 20 $ at the agreement, and we are to make it up 200 $ in August next, and the other 500 $ in 5 years with interest. We have agreed to be cooperative that is to have equal shares and equal wages, and if one should lay down more money than another he is to have interest for it. I shall now give you an

Invitation in the language of the scriptures saying come over and help us. I should like you to write immediately and tell me your mind, for if you are for coming to this country I should like you to come immediately for I expect if things go on as they do I shall either have to work for less wages or quit, and if I have to quit I expect I shall go to the west and then if you was to come we could go on to the Farm together, but remember that instead of using picks and mattocks we shall have to use axes to hew down large trees. We have just heard of a revival of trade in England and if it is so I should like my Father and you to take the opportunity of disposing of your goods and come and join us in our undertakings. I still work in the Machine Shop. I have bought some tools to the amount of about 30 dollars. I am recieving 20 $ per month which as things is at present is as much as I can expect. If I fail in giving you information you must excuse me and when you write you must ask what you want to know the worst and I will try to answer you. My Brother John is slubbing at this place. George and James H. and Joseph Kenyon and Geo. Brown is all here and in good health. I shall now ask you a few questions. I should like to know how you went on with David D Lee about the house and looms at miry lane bottom. Also I should like to know something about the upper spout how it was ordered, and how you go on with your Neighbours, and whether you can get your debts or not. Tell Thomas Levett that if he was here he might earn about 7 dollars per week with spinning and his children such as was big enough to tend the Power Looms might earn about 3 or 4 dollars per week.[14] Give my Respects to all inquiring Friends

[14]The men who operated the spinning jacks of this era had to be highly skilled; consequently they drew relatively high wages. "Tending" one or more power looms, however, was not as demanding; the young man or woman who did so was not as well paid as a spinner. Cole, *Wool Manufacture* 1: 238-239.

and tell them that I am in the Land of the Living. I remain with respect your ever loving and Affectionate Nephew

Jabez Hollingworth

[To] Mr William Rawcliff
Oldfield near Honley
near Huddersfield
Yorkshire

❧ ❧ ❧

John P. Barnett to William Rawcliff

L.pool August 11th. 1827

Sir,

I have the pleasure of Acknowledging the rect. of yours of 2nd. Ultimo, and have to inform you, that I saw your relations safe and well in Boston—— I was not Able to dispose of the Cloth in Boston at the price which you named, but took it with me to New Orleans and not finding any market for it I sent it Back to Boston, to Coolidge Poor & Head, subject to the Order of Mr Brown[15]——I have heard of the safe arrival of the vessel at Boston which I sent it by and presume it is Safe—— I am sorry I could not Sell it, as it caused me a deal of trouble to get it on and off Shore without Duties. This is all the information respecting it which I can give you—— If you Should Know of

---

[15] Samuel F. Coolidge, Benjamin Poor, and Francis C. Head were located at 67 and 69 Kilby Street in Boston. Their firm purchased the raw material and sold the finished product for the Saxon and Leicester Factory. *The Boston Annual Advertiser, annexed to the Boston Directory* (Boston, 1827), p. 72.

any Persons wishing to go to America, I shall be much obliged if you will Send them to me at Bangor. I Shall probably sail from Bangor to Boston near About the 1st of september. If you Should have Occassion to write me, Direct to Captn J. P. Barnett Brig Plato, Care

of Messrs Saml Worthington & Son

Bangor

Your Obdt Servt,
John P. Barnett

To Mr Wm Rawcliff

⁂

J. Wadsworth to William Rawcliff[16]

Pokeepsie. March 7th 1828

Dear Cousin

I received a letter from you some Months ago desiring me to inform you how my Cousin Joseph Hirst was going on – and that you had heard a bad Account of him. Truth is a verry scarce Article in all countrys, but I have found more of it here than in England except among English men. They mostly remain the same transplant them where you will and it seems that they will export lies rather than truth of which you are in want.

---

[16]Wadsworth was the owner of a woolen factory where, beginning in 1826, he manufactured broadcloth and cassimeres. Williams, *New York Annual Register,* p. 156; U.S., Treasury Department, *Documents Relative to the Manufactures in the United States,* 22d Cong., 1st sess., 1833, House Executive Document 308, vol. 2, pp. 64-67. For a contemporary view, see John W[arner] Barber and Henry Howe, *Historical Collections Of The State Of New York . . .* (New York, 1841), facing p. 141.

I have the Pleasure to inform you that Joseph is verry well in health and doing well. If to be diligent at work earning all he can and keeping what he earns is doing right then he is your man. I find no fault with him because I see none. We have no one steadyer about the place.

But I think it would be better for him to get his family here too. But he seems displeased with something. But he frequently talks of them and wishes they was here.

Give my best respects to Your Father in law and all the family. I wish they was here in this happy Country which after all some are dissatisfied with—but I believe not one in this place would exchange it for England. Joseph is going to write you next week and will tell you all about

Yours truly
J. Wadsworth

[To] Mr William Rawcliffe
Oldfield-Honly
Huddersfield
Yorkshire
England

❧ ❧ ❧

Joseph Hollingworth[17]
to William Rawcliff

South Leicester. Tuesday Morning May 20th. 1828

Dear Uncle and Dear Aunt
I'm very glad to say
That your kind letter I received
On the fifteenth day of May.

It was Dated March 25th. We was short of provision on board the ship or at least it was not the right sort. We had plenty of

[17] The author emigrated a few months before he wrote this letter.

buisquit enough for 3 voyages. We could not eat it. I believe I ate as much of it myself as all the rest. I contrived to pound it in a bag (made of sail cloth for the purpose) and made it into Pudings. That was the only way we could eat it. We bought 20 lb. of flour in addition to what we had when you left. Some of our provision was good but we had some not fit to eat. We ate the good first and we had finished a great part in the first 4 weeks. Our potatoes would have lasted out pretty well had it not been for a set of dishonest rascals. I mean the paddys.[18] We had nearly 1 third stolen, but it will take to much of my time and paper to give you every particular relating to this circumstance. Sufice it to say that should you ever come to America or have to buy provision for any body that is coming O BEWARE of BECKET! That Infernal wrech who when he could subsist no longer by riding his own country-men in Irland came over to Liverpool to impose on my Honest Countrymen who are flying from the wrath to come and going to seek an asylum in a country where that Villanous BECKET would be brought to Justice.[19] The most appropriate punishment that could be inflicted on that imposter would be to confine him in the Middle of the Atlantic Ocean in old Isaac Hicks and feed him on his own Biskit and stinking water but to conclude, O Beware of Becket.

I saw Joseph Hirst when I was at Poughkeepsie. He was a Spiner at Wadsworth's Factory. He was the only man that I have seen wear Breeches in America. I had no particular conversation with him. He asked me some Questions and I answered them. I have forgot what they were. My Mother told him that his Wife was badly situated that she wanted to come to Ameica

---

[18] "Paddys" was a contemporary pejorative term for Irishmen, according to the *Oxford English Dictionary* (Reissue of 1933).

[19] Becket was apparently an Irishman who sold provisions in Liverpool to unsuspecting emigrants and cheated them in the process.

and soon. By what he said then and by what another person told me, on whose word I can rely, I got to understand that he has no intention for sending for his wife. Mary Hollingworth knew her Grandfather when she first saw him here. She did not know me at first but when I began to ask her about the Oldfield she could recolect both you and your Daughter Mary-Ann. She says if Aunt Nancy was here she would say "Thank you mam" and give her a cent for the frock. She goes to school and seldom failes to bring home a ticket for reward of [. . .] I don't know what reason Jack-o-Micks has to give such a bad acount of America. I dare say he'll not find a Factory in England where the workmen will subcribe him Mony to send him back to America. I would scorn say anything about such a Black-Gard as Haddock for I don't calculate of fighting him with my hands but should not mind doing it with my tongue provided he would not tell so many Infernal lies. William Perken knows very little about America or the American Manufactory. He only came from New York to this place and I believe took the same rout back-again. 'Tis true that they work on sundays here. Bro' Jbz. has to work every 2 or 3 sundays in the factory repairing macheniry and doing such work as can not be conveniently done on other days. Bro. James is a spiner and he was odered by the Bos spiner one sunday afternoon in the Church while attending Devine Service to go spin that evening soon as the church service ended, but he neither woad nor did obey. I had to go tenter the second sunday after I began work and was ordered to go again but I did not obey and I have not been on a sunday since.[20] The Factory system is the caus of this.

---

[20]Tentering is one of the finishing processes. At the time, cloth was held in place on a frame (by tenterhooks) for drying outdoors; since it was stretched between the hooks, the cloth dried without wrinkles and to a uniform width. Asa Ellis, *The Country Dyer's Assistant* (Brookfield, Mass., 1798), p. 109.

I hate to see a factory stand
In any part of the kown land
To me it talks of wickedness
Of Families that's in destress
Of Tyrany and much extortion
And of slavery a portion
I wish that I no more might see
Another woollen Factory.

John says he will write you in a few weeks. Jbz. says he would write again but that he has told you all he knows, he has nothing to wright, you must excuse him. My Mother was a little sea sick on our voyage but it cured her leg. Sister H. was never sick at all but she could not walk when we landed. Mother likes country "vary weel" but she has got disapointed she has got more work than ever [in] this a land of labour. Sister H. knew her Father as soon as she saw him. He took her into his arms and she would not leave him for several hours. On the 2nd of March Miss Hollingworth was brought to bed of an Anglo-yankee. The Black Cloth which you sent Capt Barnet came here in Febry.[21] It was 12 yds long. You must write when you get this and tell me how you get along with regard to your soar hole and other matters relating to that concern and you oblige your Inteligent

Joseph Hollingworth

N.B. Please to tell Thomas Crooks that George Mellor wishes him to keep on his sick Club at Nether Thong.
Give my love to your Mother Brothers, Sisters, Samul. Wadsworth, and Jonas Brook of Westwoods.

---

[21]This may have been the same material that Americans called "negro" cloth, although the context in which it is mentioned leaves me uncertain. See Cole, *Wool Manufacture* 1: 201.

Always when you send a Letter put in a few garden seeds of some Kind. In your next a few [. . .] spring seeds.

P.S. When you this letter have reciev'd
Its contents read and all believ'd
Then put it in your Chest
Don't let each Busibody view
What I have writ tho' all is true
They would begin to Jest. – J.H.

You may let Mr Clay see this letter if you like.

[To] Mr William Rawcliff,
Oldfield Near Honley,
Near Huddersfield;
Yorkshire: England.

By Packet Ship to Liverpool

❧ ❧ ❧

George Hollingworth to William Rawcliff

South Leicester June 28th 1828

Dear Brother

I have long had an intention of writing to you and giving you all the information respecting this Country I possably could, but has hitherto been detered by a perplexed mind, for which neglect I must beg your pardon. I must begin by informing you that we are all at present in good health with the exception of Son John who has not been well for these few weeks past, but perhaps is now a little better. With respect to the Weather they say the past Winter as been verry mild. At any rate it has been verry fare from being insuportable. I have experienced colder Weather in England then I have yet done in America. The truth is that it will freeze ten times as quick and ten times harder then it will do in England and yet will not feel any colder. With respect to our Summer Weather it has hitherto been verry pleasant. It has been uniformly warm with now and then an hotter day. We are almost allways at these times favioured with a clear and pure atmosphere and an exhillerating breeze which enables us to sustain the heat of the day with verry little inconveniance. The prospect of the Country at this time is beautifull behond discription, the Woods and Fields ornamented with beautiful Flowering Shrubs of almost every discription such as are considered rare ornaments in your Gentlemen's Gardens in England. Vegetation here growes most rapidly and luxuerantly. You would be surprised to see the uncultivated state of the Land and see the abundance that it will produce conected with the little labour that is bestowed. We have planted about half a Road of Potatoes in verry rough Ground without any manure and they say they will grow verry well. Whether they will or not is yet to prove. Of this I am certain that had they been planted in like manner and with as little labour in your part of England that they would produce nothing. We have made a verry neat little Garden attached to our house in which we have now growing cucumbers Melons, Squash, various kinds of Beans such as is not known in your Country, Carriots Lettuce Parsley Beets Marygolds Sweet Williams Saffron Sage Hysop and various other things all of which we are raising from the Seeds. The uncultivated land even the Road Sides produces abundance of Red and White Clover and are excellant good Pasturage. The horned Cattle are almost universelly of one Kind and Collor either Lighter or Darker Branded or Red. A great many of them appears to be good ones but very little care taken of them. The Yankee generally takes a

Vast more care of his Horse then any thing else, and they are worthy of care For the breed of Horses here are generally the most fleet and Active I ever yet Saw. I had sufficent proof of their agility in our Stage rout from Albany to Leicester a distance of 124 miles and some of the Road the most teriffic for a Stage road I ever saw. The Stage drivers are super Excellant at their business. When I was at Poughkeepsie I was acused by Joseph Hirst of having brought a letter from you to Mr Wadsworth enquiring after his conduct. I told him I had not brought any letter for any Person and that if Mr Wadsworth had received a letter from you J. Brook most probably had been the Bearer. He appeared to me to be offended at this and other things which had accured. He told me he had fully expected you and Joseph Drake following after him to America but that now he perceived you were grown Cold upon the subject, and that by recent letters received he discovered that you almost discreditted his report of America. This said he, had almost fully determined him never to write to Old England while he lived, and that if he had his Children here he would not truble himself about any other person. This is what I know respecting J. Hirst. Which I beg of you to keep very SNUGG. We all live together in a double House.[22] We have plenty of room. The House contains 8 Rooms besides a Celler under the whole. We pay 60 Dollars a year Rent. Geo. Mellor James Hollingworth and & J. Kenyon Boards with us. Son James is a Filling Spiner viz. a bobing Spiner and can earn 8 or 9 dollars per week.[23] Son

---

[22]Apparently the "Rhode Island" system, in which whole families were hired, was being used in South Leicester to recruit a labor force. For a brief but lucid discussion of this and the "Waltham" system, see Edward C. Kirkland, *A History of American Economic Life*, rev. ed. (New York: Appleton-Century-Crofts, Inc., 1939), pp. 332-341.

[23]Terminology in the textile industry varied almost from mill to mill; it is not easy to determine just what Son James was doing. He was

John has been a little time a Condenced Spiner. This is a new meserable business for making money. He is now a Slubber which is a fare better Job. He will make better then a dollir a day.[24] Son Joseph has been a Gigger every since he came at only 15½ dollers per Month. As the Job was a new one to him and one I wished him to learn I was not particular about wages at first, but now has he has got an expert hand I am thinking of asking for more wages not less then 20 dollers per month.[25] Edwin is a Warp Winder or Spooler winder for the Warping

---

probably spinning filling (i.e., weft) yarn. This could also explain the term bobbin spinner, since weft yarn was wound onto bobbins that fit into the loom shuttles. Weft spinning was done with a different technique than warp spinning, since the warp had to have a tighter twist. Abraham Rees, "Woollen Manufacture," *The Cyclopaedia; or Universal Dictionary of Arts, Sciences and Literature*, vol. 40 (Philadelphia, 1822).

[24]Even more obscure is the task performed by John. As a condenser spinner, he may have been operating one of the many devices invented and tried in order to eliminate the time-consuming operation performed by the slubbing billy. Edmund Burke, *List of Patents for Inventions and Designs . . .* (Washington, 1847), p. 68; Cole, *Wool Manufacture*, pp. 101-102; S. N. D. North, "The New England Wool Manufacture," *Bulletin of the National Association of Wool Manufacturers* 29 (1899): 259. In 1826, John Goulding of Dedham, Massachusetts, invented a successful condenser consisting of a fluted roller that converted the carded wool into narrow, continuous ribbons, and revolving metal tubes that condensed the ribbons into unspun yarn called roving. As the pieces of roving came out of the condenser, they were wound on a spool that could be placed directly on a spinning machine. Apparently Goulding's invention had not yet been adopted in this factory.

[25]Gigging, napping, or teasling, as it is sometimes called, is a finishing process in which the surface fibers on cloth are brushed and raised. Joseph seems to have been napping by hand, although machines were in use in some mills by this time. Monte A. Calvert, "The Technology of the Woolen Cloth Finishing Industries . . ." (North Andover, 1963), pp. 41-43; Andrew Ure, *A Dictionary of Arts, Manufactures, and Mines . . .* (New York, 1845), p. 1325.

Machine at 1½ dollers per Week. I am intending to have his wages advanced also.[26] There is not much in America I dislike [excep] ting the too general conduct of Emigrants, and the Factory Sistem which Sistem I hate with a perfect hatered as being only calculated to create bad feelings bad principles and bad practices. I must inform you of a disapointment I have experienced namely when I set off or rather was forced or driven off by Family and circumstances, I was fully perswaided to believe that my nearest Relatives viz Brothers & Sisters would Speedily follow after me, but now that I am in America how woefully I am undeceived by recent letters which have been received in which the writers state their relinquishment of an intention to come to this Country, and courting their Sons to return by flatteryïng accounts of the flimsy prosperity of Old England. These circumstances fully determines me never to trust to either Person or circumstance any more, nor will I try to perswaid any person against his mind to come to this Country. If any person is inclined to come from principle (and none else are fit to come) I will most willingly give them the best information I am able. I particularly request you to write as soon as possable and therein declare your Intention whether you will come to America or not. If you should come it would be of no practical use to send you any money to England. But should you finelly determine to Stop in Old England then we will send you some Money as soon as possable. While I am writing Sons James and Joseph are Fishing in the Factory dam which is a Pond of Watter containing A surface of 40 or 50 acrs.

---

[26]As a warp winder Edwin was responsible for transferring yarn from spindle bobbins to the spools that fitted in the warping creel (a rack for holding the spools). The warping reel then "dressed" the warp threads, i.e., drew them parallel and added a certain amount of tension as they were wound around it.

Son John's Wife was brought to bed some Months since of a verry fine daughter which they call Elizebeth – We all unite in our Special Regards to you and your Wife – and am Dr. Bro.

Yours affectionately
Geo. Hollingworth

If you think proper you may lett my Brother John see this Letter.
I had forgot to mention son Jabez. He is working in the machine Shop at 25 dollors per month.

[To] Mr William Rawcliffe
Oldfield Honley
near Huddersfield
Yorkshire
Old England

❧ ❧ ❧

Joseph Hollingworth
to William Rawcliff

South Leicester, September, 21st. 1828.
Honley Feast.

Respected Aunt and Uncle

'Twas on the 6th day of September,
A letter came, I well remember,
From "Uncle William,, unto I;
Dated the 6th day of July.

On the 4th. of July I wrote a letter to you but never sent it. I give you an extract

I'm inclin'd to write to you this day,
Nothing shall cause me to delay;
[. . .] iring, that you will keep quite,
[. . .] Sentence I shall herein write. –
[. . .] oud Tyrants of European Blood,
[. . .] Independence Shew'd his face
[. . .] Liberty her graces good,)
Did seek This Country to abase. –
Each Foreigner that's in this Land,
Natives likewise Join hand in hand;
Celebrate at your own homes,
Each Independant day that comes." –

"This day is the 52nd Anniversary of the American Independence, and is celebrated with enthusiasm by the most zealous Patriotic Citezens thr'out the United States. At this place the day opened with the confused ringing of the 2 factory Bells & church Bell. They had 3 cannons which they kept firing at intervals thro'out the day. There is a pond at this place wich covers a surface of about 40. or 50 Acers in which are situated two Ilands. On one of these a Bowling Ally was erected and those people who had a mind to, went over to play at 10 pins. The Respectables had a diner at tavern & I am sorry to say that some instead of celebrating the day aright gave way to Intemperance & woshiped Baccheus.[27] None of our family went either to the Dinner or Bowling Scrape. We celebrated the Independence at home by drinking each of us a toast. I repeated the following, tho' not thought of before"

---

[27] A contemporary drawing (Plate I) suggests that Fourth of July celebrations in Worcester County were not quite as disorderly as the author suggests.

"May Nations all, both great and small,
True Liberty enjoy;
I'm for that cause, and ever was,
E'er since I was a boy. –"

I have seen plenty of stone rocks in America. They are not all wood ones. When you come to this country beware of Becket & Scholfield.

In answer to your N°. 2nd. I must tell you that I like the Indian Corn Porridg exceedingly well, much before oatmeal Porridg, it being of a richer quality and more wholsom. You say the Compliment I sent in my last for Thos. Crooks sounds like coming back, wich you call Bad News! Now I feel almost at a loss what to write lest I should send you BAD NEWS. If I was to write a Letter full of Nonsense representing America ten times better than it realy is or better than any Country on the face of the earth could be, If I was to compare it with Elysum to tell you the Natives were Angles that we could have Gold for gathering, that all Wimsiecal and discontented Englishmen that came to this country were content and happy and that not one of them ever thout, of returning home, no Never! Never! Never! If I was to fill a sheet of Paper with such like stuff as this, Then you might, Parhaps you would, Call it good News. But far be it from me to write any thing but the truth, for what News can be better than the truth? George Mellor says that if he was as young as he is old he would not return Back at all.

I am sorry to hear of the Death of Allen not that I cared any more about [. . .] his Father but by his Death the Father will gain more Life. I wish [to know] how Jabez got the name of sexton. Should your next Anounce the Death of St. Rachel I shall feel a little glad for the good riddance the neighbourhood will have of her. You used to say, that if I came to [this coun] try, you was afraid I should turn Drunkard. Now let me tell [you] that I have not spent one cent in liquors scince I landed in America. Before I had been in America one month I could with truth say I had drunk' more Spirutus Liquors than I ever did in England all my life! And yet I was never intoxicated tho' I might [. . .] been many-a-time at Poughkeepsie and elsewhere. Although [we were] at Poughkeepsie nearly 2 weeks & Mr. Bowers & Mr. Brook [lived] about as near together as they did in England; yet for all that my [mo]ther has never seen Grace Brook in this Country!

When I left England I anticipated happiness which gave me great encouragement. I thou't of going to a good land where I should meet with relations who had gone before me & those whome I left behind I expected shortly to follow. In my anticipations I have not yet been decieved and wether I shall partly be decieved or not it is in part for you to determine. It is a very great undertaking to be sure for People to sell all theire Property to leave the land of their Nativity to cross an Ocean more than 3000. miles wide and go to a land fare away amongest Strangers, but when we come to consider of the wellbeing of ourselves and of our Posterity we ought to shake of our fears and to act with an undaunted courage and resolution for though I have mentioned the extent of an Ocean 3000. miles yet it was not with an intention to cause dismay, for I think thus and hope you will think the same

If I'm destin'd to a watery grave
There's none on earth that can me save
If I'm to die upon dry ground
There's none on earth that can me drown'd

Septbr. 22nd. Honley feast, Monday Night. 7. oclock. I guess you have been chusing surveyers to day. I will write a little more bad News. i.e. we have taxes to pay. What! taxes in America! Yes we have to pay two a year, the first called "Road & County tax."

Plate I

The Southbridge Light Infantry, on the Common, 1826. Pen and water color, by C. L. Ammidown. Gift of Edgar W. and Bernice C. Garbisch. Courtesy of the Columbus Gallery of Fine Arts, Columbus, Ohio.

You will guess the use of this tax by its name so no more of it. The second is called "School tax." You will not perceive the use of this so readily as the former but I will try to explain it a little. This great Commonwealth is devided into 24. states, these states are again devided into countys, counties into towns & towns into School Departments and it is enacted by the Laws of the United States that the Farmers [in] every town shall meet to levvy a tax, yearly, on all the inhabitants above the age of 18. That the Money raised in every town shall be [divid] ed eaqualy amongst the schools Departments according to number [of] scholars in each Department. All persons under the age of 21. are scholars and may go to school if they please. I when in my own Native [land] if I went to school I must pay for it, but here I can go free.[28]

Honley Feast, Wedensday Night. 9. oclock. We have begun waking [. . .] the Factory tonight & shall have to continue till about 1st. of March. May God Blast the Factory Sistem altogether for what I care about it. We leave work at night by Moon Light & returns in the morn' by the same.

Sat. Septbr. 27th. Flour here is 9 dollars per barrel. I heard a few months ago that John Woodhead has got Married and was living at New York. He came on the same ship with us & she did the same. She is a Manks Girl. Joe Fletcher was buried the very day we arived at Poughkeepsie. Bro. James was at his funaral. I heard his Funaral Sermon preached at the Methodist Chappel on the Sunday after. At Pleasant Valey I saw John Scholfield John Boothroid & Mick Boothroid. They all lived at Pine Grove

---

[28]Joseph's civics lesson was not entirely free from error. Of course it was not federal, but state law that governed local school districts. Persons above 16 years of age, not, as Joseph claims, 18, were subject to poll taxes in Massachusetts. *Massachusetts Laws* 1805, c. 119 § 2D.

1 mile from Plasant Valey.[29] John Boothroid had a ten weeks voyage and when they arived they had neither masts, yards, sails nor riging left. I heard that John Hobson has got Maried and keeps a tavern on Long Island. At Wocester 7 miles from here is a new Coal mine. The proprietors have advertised for 15 miners.[30] Co^{sn} J. Hollingworth has this week recieved a letter from Uncle John. With respect to that peice of land at the back of your house George Mellor says that John Haigh claimed it ever scinc he knew it & how it came to be his he knows nothing about it. I say that where I in your place I would not care any thing about it. Why should you be bothering about a yard or 2 of land wich (if you should get possesion of it) wil not be your own, when you might in this Country get a comfortable home with 50. or 60. acers of good Land, and none would molest you or dispute the right of your Soarhole or back window. Excuse me for not sending you a whole sheet paper. This was the only one I had when I begun writin.

And so by the Blessing of God I remain your Affectionate and Inteligent Nephew

Joseph Hollingworth

---

[29]Home of the Pine Grove Woollen Manufacturing Company, which produced broadcloth, flannel, baize, satinet, cassinets, kerseys, and narrow cloth. U.S., Treasury Department, *Documents Relative to Manufactures*, p. 90. The John Scholfield referred to here is apparently a recent immigrant and not a member of the family associated with the wool industry in Massachusetts and Connecticut. For an assessment of their contribution, see Grace L. Rogers, "The Scholfield Wool-Carding Machine," in *Contributions from the Museum of History and Technology* (Washington, D. C.: Smithsonian Institution, 1959), pp. 1-14.

[30]Colonel Amos Binney worked an anthracite mine there for a brief period beginning in 1826. William Lincoln, *History of Worcester. . .* (Worcester, 1837), p. 354.

My respects to all enquiring Freinds and to yourself. Tell Jonathan there are plenty of Girls in this Country. Some have Larg fortunes.
Father sent you a letter about two months ago. Tell me if you have heard of John Evans scince I left. [. . .] you can & direct as before.

N.B. The Day you Dated your letter was your Birthday & the Day I recieved it was my Mothers Birth Day.

[To] M^{r}. William Rawcliffe
Oldfield Honley
near Huddersfield
Yorkshire
Old England

❧ ❧ ❧

Joseph Hollingworth
to William Rawcliff

N°. 4. South Leicester, Dec^{br}. 7th. (One year old My New Birth Day. [1828]

Respected Aunt & Uncle

In England I've left, both relations and friends,
Whome I respect more, than just for my ends;
For If I did not, than no more would I write,
Since they are far from me, and out of my sight.

Don't be astonished at my writing N°. 4. before I have received N° 3. from you. One reason is this, viz. that this day being the 1st. Aniversary of my landing in America I wanted to celebrate it by writing a Letter. Father sent you a letter many a months since but has had no answer. I am afraid you did not get it. Bro. John sent you a letter 3 weeks ago (by Joseph Brook who is returning to England) wherin he states "that he & James Hollingworth was going to take a small Manufactorien place in Herkimer County, New York."[31] But they have not taken the place on acount of the Owner wanting too much Rent. At present they are working at Winstead in Conecticut.[32] Joshua Smith is at the same place. If you write to Bro. John you must not Direct to any place but South Leicester untill you have had another letter from him.

I understand that Johnathan Hirst has arived in this Country and that he is at Poughkeepsie. Also that he and you & the rest had a Lawsuit with John Hirst about your Fathers Property. When I received your last letter I anticipated somthing of this kind, because you said you was at Halifax on the 12th of July and the letter bore the Halifax post mark of the 19th of July which made me certain that either you or some body else was at Halifax a second time for what reason I could not tell but was afraid it was about some law scrape. I thought it might be with Messrs. Lees though I hoped not because in Law Scrapes winers are losers.

I have lived in America exactly one year. I have seen all the Seasons and must confess that I prefer the American weather far before the English. I have never seen in this Country a

---

[31]Apparently there were several small cotton and woolen mills in the county. See Williams, *New York Annual Register*, pp. 149, 157.

[32]Winsted was a village inside the township of Winchester. The Hollingworths probably worked for Alfred French, who made broadcloth there. The village also included a dye goods dealer and a shop where "Machinery for Woolen Factories" was made. *The American Advertising Directory For Manufactures and Dealers in American Goods* (New York, 1831), p. 147; for a contemporary view, see John Warner Barber, *Connecticut Historical Collections* . . . (New Haven, 1838), p. 503.

Beggar such as I used Daily to see in England, nor a tax gatherer with his Red Book as Impudent as the D-v-l, taking the last penny out of the poor Mans Pocket. In this country are no Lords, nor Dukes, nor Counts, nor Marquises, nor Earls, no Royal Family to support nor no King. The "President of the United States" is the highest Titled fellow in this Country. He is chosen by the People, out of the People; holds his station four years, and if not rechosen he is no more than the rest of the People. The President when he makes a speech does not begin with "My Lords and Gentlemen" but with "Fellow Citizens." When we came from Poughkeepsie to this place we stopped the first night at Amenia [N.Y.] . I was astonished to see the Driver lose the Horses off and leave the waggon by the road side. I spoke to Bro. John and told him I thout our goods would be stolen. He answered "No Danger." If I had not been well tired I believe I should not have slept that Night for fear, but I no ocasion to fear, for our goods was as safe as when locked up in the yard at Manchester, and all the journey thro' we never put our waggon under cover, but always left it in the road or street (or where ever it happend to be) all night to the Mercy of the Theives, if any there were. We have no lock to our doors, we never mak them fast at night. This helps to confirm the truth of some of my old English Poetry

> A land where tyrany is no more
> Where we can all be free
> And men without a lock to th' door
> Sleeps in tranquility

N.B. There is the Factory Sistem which breeds a kind of petty Tyrany but ere long will be leveled as low as its supporters I Hope.

Give my respects to Old Haigh and tell him if you please that my Father has no occasion to hawk Nuts in America as every body can have them for gathering in this Country. Neither is he bound to carry Mes^{rs}. Haighs wet Peices up Mirylane on his Back nor to go Roast himself in their stove every Sunday morning for Nothing. I have got a New Hat which cost 5 dollars and three quarters. I have a pair of Boots making which will be 4 dollars and an half. I am still working in the Gig Room at 17 $ per month.[33] Father has been writing a letter for 6 months together to William Lockwood. He has not yet finished. It is to be so large, and so compleat with information, but I guess it will be like the Mountain in Labour, it may bring forth a Mouse. The more I live in this Country the better I like it. Please to get that book called "Jamisons Atlass"[34] out of Meltham Club and bring it me when you come to this Country and I will pay you whatever it costs. You must excuse my Brothers James and Jabez for not writing as they are both deeply engaged in Sparking. Jabez Sparks a yankee Girl James Sparks a Saddleworth Girl, and on the 25th. of Nov^{br}. Joseph Kenyon took two English Girls to a Ball.

We are all in good Health at present hoping you are the same. Jabez & James are a little tickled at what I have Just written So I will conclude.

I Remain your most Inteligent Affectionate &
well Wishing Nephew
Joseph Hollingworth

[To] Mr William Rawcliffe
Oldfield Honly
near Huddersfield
Yorkshire
Old England

By Packet Ship from New York to Liverpool

---

[33] See George Hollingworth to William Rawcliff, June 28, 1828.

[34] He was referring to Alexander Jamieson's *Celestial Atlas* (London, 1822); Joseph was apparently an amateur astronomer.

Joseph Hollingworth
to William Rawcliff

South Leicester, Feb. 8th 1829

Most Respected Aunt & Uncle

I received your Letter, Nos 3. Janry 22nd Excuse my not writing sooner as it not so much through neglect as other causes. I now am a Shearer and we have had to run the Shears Night and Day.[35] Of course I wanted to rest on Sundays. Other causes I forbear to mention. I think I percieve by your Letter that you have a wish and even an intention to come to this Country the ensuing Spring. I hope you will, and I can assure you that tho' you might find some difficulty or perhaps some little disapointment at first yet after all when you get a little Settled and See the good effects of your past toil and exertions you will never rue the day that you left Old England. But you say if you was in America What could you do, as you would be clumsy at weaveing or Spining &c. Now I will tell you as plain and in as few words as possible. When you land at New York, amongst other fine things, most probably you will see larg Boats of 300. Tons burden, loaded with FAT HOGS! all killed and piled one on another. You will naturly be led to Enquire, Where do all these Hogs come from? They are brought down the North River. And up the North River I would have you come till you arrive at a place called Poughkeepsie. When you arrive there you will soon find what to do. You may take an House in the Town

[35]Shearing is the process by which the raised nap on woolen cloth is cut off. The implication is that shearing was done by a rotary shearing machine, which superseded the hand shears (Plate II) with which the Hollingworths were probably familiar.

Plate II

Shearing in a Huddersfield shop, ca. 1815. In the *History of the Huddersfield Woollen Industry* by W. B. Crump and Gertrude Ghorbal (Huddersfield, 1935). Courtesy of the Tolson Memorial Museum, Huddersfield, England.

for your family till you can Seek out a better Situation in the country, or you may keep a Store or a tavern or whatever els best suits. Joseph & Johnathan Hirst will be better able to give Information on these subjects than I can. I am sorry to hear that you still continu to have such voilent pains, but more probably a Sea Voyage would do you a deal of good. I Should advise you to Settle all your concerns to Sell all your Property to get in all dues and to pay all Depts if possible, and when you get to this Country you will be as one that is Born again. Concerning the Money which Father, Bro. John & Jabez owes you I have this to Say, that I have heard them say that they did not like to run the risk of sending money to England but if you come to this Country they will pay it you the first oppertunity. Your Cousin John Woodhead is now prepared for two worlds, as he has got a Wife in each of them. Does is Wife in the old World know that he married a Girl at New York, and after living with her some time at that place he took her up the North River and up the Great Westren Canal to Buffalo (a town on the borders of Lake Erie) and there left her? Tell Uncle John Hollingworth that his Son James has got married to Seraph Livermore the Daughter of Salem Livermore, who owns upwards of 600 acers of land in this town, and for whome Salem Woodward (the Captain of Isaac Hicks) used to work when a boy. I am in good health, and altho' I [. . .] up close in the Factory working Nigt & Day and tho' I think and study and dream a great deal about you yet, I keep up my Spirits and my face is as big and as red as ever. Wishing you Health and Prosperity I Still Remain you Affectionate Nephew

Joseph Hollingworth

P.S. I cannot conclude this Letter without saying somthing of buisness that is not altogether mine. It has been agree'd on here by J. Kenyon, J. Hollingworth &c. that the best plan is for you, Uncle John Hollingworth, & Uncle John Kenyon, and Families, (should you all be disposed to come to this Country) to come together, and that you should assist them a little with money should they not be able to make out for themselves, and whatever expences you incur on their account Joseph Kenyon and James Hollingworth will pay as soon as you arive in this Country. On the above plan I have a few Observations to make. 1st. you have already done a great deal for persons that has come to this Country. When I used to fodder the Cattle I always pulled on that part of the mow where It came the best. But, 2nd Cousin James Hollingworth is a good fellow, he will do what he says. I dare be bound for him. He has behaved very well to Brother John and he assisted us with 60 dollars when we came to this place at first. Should you be able and think fit to assist them if they need any assistanc I should advise you to do it on uncle J. Hollingworths account and have suretys for the same. Then there will be no fear but you will be Recompensed. "Begin nothing of wich thou has not well considdered the end". "Blessed is the man that doeth good.

> I. Heap may tell tales, Haigh & Wilson may scorn
> But tell them poor fellows I am better Born
> Than to heed or to care, for ought they can say
> Since I am living well in America.

Be not affraid of geting something to do, I have not wove a pick scince I came here,[36] yet I have had work enough work that I prefer before weaving.

Joseph Hollingworth

---

[36]A "pick" in textile terminology refers to one pass of the weft through the opening formed when some of the warp threads on the loom are raised and the remainder lowered.

March 7th. 1829

Farewell dear friends, Farewell, Adieu
Until I meet again with you
In this great land fertile and good
Abounding plentiful in wood

But if we never meet again
In this dark world of greife and pain
May we assemble at Gods feet
Here happiness Shall be complete.

[To] Mr. William Rawcliffe
Oldfield Honley
near Huddersfield
Yorkshire
Old England

By Packet from New York to Liverpool

⁂ ⁂ ⁂

Joseph and Jabez Hollingworth to William Rawcliff

South Leicester Sept. 6th. 1829

Dear Uncle and Aunt,

To leave old England you've contriv'd,
The dangers of the Sea survived,
And in this land have safe arived;
I heard with pleasure.

But uncle, why are you so quite?
What is the reason you dont write?
A letter would give me delight;
And be a treasure.

On the 10th of august I recieved youre last letter informing me of your intentions of sailing on the 8th of July. J. Kenyon, as I understand, has received a letter from J. Brooks, at Middletown in Connecticut, which states that he, you, and several others landed at New York on the 17. Ultimo. that Cousin Mary Kenyon had come along with you, partly at your expense, and that you had behaved to her like a VILLIAN all the way on her passage. &C. &C. I neither saw this letter nor even so much as heard it read, but what I have stated I heard from Kenyons own mouth, or somthing to the same purport. For my own parte I did not believe all this, because I never knew you behave like a villian to any body that behaved well to you. (You may think this is flattery.) So I concluded patiently to waite for your own account of things and circumstances. I have waited now till my patience is exausted and I can waite no longer. I have heard it rummourd about that J. Kenyon has had a letter from you, but he keeps it so snug that I cant get to know a word of its contents. I do not know when he recieved nor any thing more about it.

We are all in good health at present. I sent you one letter to England in July so that I guess it would pass you on the Atlantic. Bro. John lives at Oxford 6 miles from here. He took a Farm at that place last May of about 80 acers rent about 50 dollars per annum.[37] He and family I believe are in good health. I wrote to Jonathan Hirst last spring but have not had any

[37]The contemporary New England farm (Plate III) could be operated with several strong backs and a few basic skills.

answer, should like to know the reason. We all of us here would be very glad to see you all but it is expensive traveling. Besides if any of us was to come over it would endanger the loss of our work here. If you could make it convenient to come over and see us we should take it as a great favour. Besides, you might the more easier settle accounts with Father, Bro^{thr} John and Jabez. I be very glad to come over and live with you at Poughkeepsie if I could have work at the same buisness I am working at here. Please to write and give us some account of your voyage and if you have got work and of all others things &C. &C. &C.

And you will much obliege,
Dear Uncle and Aunt,
Your most Affectionate Nephew

Joseph Hollingworth

[P.S.]

Dear Uncle

Accept a few lines from one who has not forgot you nor your kindness to me. Although I have been negligent in writing to you yet it was not without Some reasons which I shall explain to you at some future period. I was glad when I heard of your intentions of leaving Old England. But my gladness was multiplied Tenfold when I heard of your Safe arrival in America. I should like to see you at South Leicester as soon as you can make it convenient; if you do not come soon I should like you to send me a letter and ask such questions as you think proper and I will try to answer them to the best of my abilities.

[Jabez Hollingworth]

Plate III

Dennison Hill, Southbridge, Mass., ca. 1825. Oil, artist unknown. Courtesy of the National Gallery of Art, Washington, D.C.

George Hollingworth to William Rawcliff

South Leicester Oct 21st 1829

Dear Brother

We duly received yours by Mary who arrived here on Saturday last. She stoped at our house during the day. At night her Brother fetched her Trunk from the Taveren and contrary to her expectations carryed it to Nephew James' House and she had to follow it to be Boarded there without knowing why or wherefore. He has not got her any Work at the Factory yet, and whether he will be able to do this is as yet a matter of doubt for things are here in a very curious and precarious state. Mr Anderton has cut and run from South Leicester and left his Wife and all his personal debts unpaid.[38] Geo Mellor has a promisory Note against him of more then an Hundred Dollars which he lent him. It is expected the Factory will have to stop when we have worked up the present Stock which will only last us about 3 Weeks or a Month. Some are of opinion it will not stop long, only while a new Company takes hold but this to us who know nothing about it is doubtful. It is also surmised that when the new Company is formed that they will apoint a

---

[38]In 1820 Anderton owned a mill in the Cherry Valley section of Leicester, where he reported the following machinery for census-takers: 1 broad shearing machine, 2 narrow shearing machines, 3 broad looms, 4 narrow looms, 2 jennies, 1 billy, 1 picker, 2 double carding machines, and 110 spindles. He sold this mill in 1825. "Questions to be addressed to the Persons Concerned in Manufacturing Establishments . . . ," manuscript returns, Fourth Census (1820), National Archives. Selected Xerox copies in Merrimack Valley Textile Museum; *History of Worcester County, Massachusetts* . . . (Boston, 1879), 1: 631.

Yankee Superintendent who will show neither Mercy nor favour to Old Country men.[39] If the Factory should Stop or any other unpleasant thing should happen so as we should have to Quit, we think to try if possoble to do somthing for ourselves. We have had some talk respecting taking that small Factory which stands by the road a little more then half way betwixt here and Son John's but upon second thought I am affraid that the engaugement would be more then we could manage at present for I understand there is no machinery in the Factory except a Water Wheel a Fulling Stock drum, &c.[40] If we were to engauge it we should want 2 Carding Machines one Billy 2 Jennies 4 Sattenett Power Looms and one Hand Broad Loom &c.[41] These

---

[39]English laborers, apparently, were not universally in favor among American mill owners: James Kempson of Philadelphia, a cotton manufacturer, contended they were undesirable because "...they are so dissipated and so discontented.... Our own workmen are better educated, and more intelligent, and more moral, and refrain from sensual indulgence." From the Factory Commissioners' Report in Great Britain, quoted in the *Mechanics Magazine and Register of Inventions and Improvements* 3 (January-June 1834): 33.

[40]He may be referring to what was known locally as the Sigourney Mill, which Joseph Stone operated as a carding and fulling mill until 1820 or 1822 (he also ground scythes there); between 1822 and 1823 he leased it to a man who introduced six or eight cotton power looms; in 1823 others leased it, and between 1826 and 1828 someone else leased the mill. The mill was closed between the autumn of 1828 and the spring of 1830. On the other hand, he may mean the Old Huguenot Mill, a cloth-finishing shop that changed hands four times between 1820 and 1829. George F. Daniels, *History of the Town of Oxford, Massachusetts* (Oxford, 1892), pp. 188-214; and *History of Worcester County* 2: 182-185.

[41]The water wheel to which George Hollingworth refers was the traditional means of supplying power to grist mills, saw mills, and fulling mills. The fulling stock drum was simply an extension of the horizontal shaft of the water wheel. The shaft extension was fitted with tappet arms, which lifted beaters (or "mallets") as the wheel turned, allowing the beaters to

would cost a great deal of Money. I am told we could buy these things on Credit, but I do not aprove on much Credit for it is the verry cause of many failurues. [This is] somthing we must do and I am verry anxiouse that we do it to purpose but it takes some time to become aquainted with all the NICK-NACKS of a new Country. Son Jabez is anxious to know whether he will be of any servise to you as he almoste expects to be at liberty soon. He is told that the Company intends to employ only one Machinist. This must be a man that can work boath in Wood and Iron. This is rarely to be found except in a Yankee who professes to do every thing. We should be verry thankful for your immediate advice upon these important subjects. I am of opinion it would be worse than foolishness to envolve ourselves in difficulties except it were in the attainment of a permanent residence and even this should be set about with great caution and deliberation. Previouse to Land buying a Foreigner should declare his Intentions and become Naturelized as soon as possable. If he does not do this he cannot during his lifetime either sell or give it to give a good title, nor if he should die can his Family heir it. I am informed that it will by Law revert back to Goverment. I am also informed there are many persons who are verry anxiouse to sell Land and dupe people of their Money who cannot give a good title. Therefore it becomes an inperative duty in Strangers to be wide-awake and to ascertain all the different facts relative to or connected with good titles of Possesion &c. Connected with this is the Quality of the Land

fall regularly on the wet cloth being fulled. The effect of this action is to shrink the cloth and make it more compact. Oliver Evans, *The Young Mill-Wright and Miller's Guide* (Philadelphia, 1829), pp. 338-339 and Plate xxiv. Satinet power looms would weave yard-wide fabrics, generally with a cotton warp and a wool filling. The hand broad loom would weave a fabric about 2½ yards wide. Cole, *Wool Manufacture* 1: 120-127.

and Coquality of the Situation. Some persons are of opinion that the Quality of the Soil is determined by the kinds of Wood growing thereon, that Pine Hemlock and other soft-woods denotes poor or Barren Soil and that Oak Maple and other Hard Woods denotes good and rich Soil, but I am at present (whether correctly or not I cannot positively tell) of a different opinion. I think the above is not an infaliable criterion by which to Judge Land. I believe I am like most other persons best able to determine the quality of land by seeing it. You inform us you have been West to see some Land there. You do not give us all the particulars respecting it we could wish. Probably hast and want of time when you wrote prevented you. The land you appear to have picked upon is some of it cleared and some of it not. but you do not say what kind of Woods is growing thereon. This is a Matterial point with respect to Value either for Fuel or Building either Houses or Machinery. You do not say whether there be any Buildings or Water previledge thereon or not, nor do you mention how fare it is from the Hudson and Delaware Canal,[42] or whether it be in the vicenity of any Coal Mine. You do not hazard any opinion respecting the healthiness of the Situation or whether you liked the general face of the Country. Upon these subjects I could wish to have particular information because if we have the idea of ever becoming Neighbours and associates we ought to confer one with another and inform each

[42]The Delaware and Hudson Canal ran from the Hudson River at Kingston, New York, to the Delaware River at Carpenter's Point, Pennsylvania, and from there to Honesdale, Pennsylvania. From there to Carbondale, the Lackawaxen Railroad provided the means for transporting coal from the mines to New York City. James Eldridge Quinlan, *History of Sullivan County* (Liberty, N.Y., 1873), pp. 655-662; H. S. Tanner, *A Brief Description of the Canals and Railroads of the United States* (Philadelphia, 1834), pp. 12-13.

other of our respective wishes and intentions that it may give Mutual approbation and Satisfaction.

I understand by Mary that Will. Lockwood is not satisfied with America. This is what I expected, and I am verry glad he has not me to blame for giving him any encouragement to come to this country. Mary tells me that M[r] Lockwood wishes me to write to him but I guess I shall defer that untill he has satisfactorly explained to me the reason why he showed my letter which I sent to him in England to Brooks Family since his arrival at Poughkeepsie. It is not the fact of Brooks seeing my letter that greives me for I dont care if Brooks were to boath see and hear every word which I have either wrote or spoak respecting them. What displeases me is the fact of Lockwoods ill Faith or breach of rules of Confidence &c. I must conclude by informing you we are in good health and we hope this will find you the same. We wish you to write as soon as possable and give us all the information and advise you can. Accept of our united regards and present the same to your Wife.

Yours affectionately
Geo. Hollingworth

Oct 29. 1829

We have heard nothing of Nephew Geo. yet, except that he is somwhere in the Neighbourhood of Philidelphia. The Man you mentioned has not come this way.

[To] M[r] William Rawcliffe
Poughkeepsie
Duches County
State of New York

Joseph Hollingworth to William Rawcliff

South Leicester Nov^{br}. 7th. 1829 Saturday evening.

Dear Aunt, and Uncle.

I write this letter INCOGNITO, I therfore desire you to keep it to your selves. I shall state a few of my grieveances in a brief manner. I should have told you somthing of it when you was here, but I thought it best not to trouble you then; besides I had no oppertunity. You well know how things went on in our Family in England, with regard to me, that I was always the BAD LAD,&c. Were I to enumerate all the evils that has been practised against me in this country, I might fill sheets upon sheets; but I forbear. Let a few suffice. You know how James and I was clothed when we left England; he had 3 or 4 good suits, & I only one, besides my old clothes, which was wore out when I got here. I had to take one of James's old coats, which has been my every-day coat ever scince; and which I have got on this present moment. I did not have any new cloths till last June, (Hat Boots & satinet trousers excepted,) and might not have had any to this day; but I told them [in] plain terms, that if they did not get me some I would look out for myself. Last winter, I had to work a good deal of overtime at Nights; had I refused the whole Family might have lost their work: so I calculated to have the overtime wages for myself. It amounted to nearly five Dollars. I have succeeded in getting 2 dol^{rs}. ONLY! which is all the pocket money I have had in this country! I might have overlooked all these circumstances, at least a little longer, had not one, transpired, of a diabolical nature. We have a new Family Bible, and I thought to make it a practice, every

Night, to read the lessons; as set aparte in the callender.[43] I did so, 3 or 4 nights when Lo, and Behold! last Tuesday night, when I got redy to read my lesson, James had locked up the Bible, which he called his; and refused to let me have it, because it was too good to be used! (Turn over

In days of old, the word of God,
By Monks, and Friars, was read o'er;
But now, a selfish PEDDY-NOD,
Has lock'd it up within his Draw'r.

This is too much for me, to have the Family Bible locked up; in a Land of Liberty and Freedom. Nor can I bear all their frumps and scornings, to be called a selfish Devil when I ask for the money which I earned when they sleeped in their beds; and ever and anon, to be told, that it is not my own Coat that I wear. I say, I can not bear it. I will not bear it. It is too much for mortal man.

I am no Cat, I am no Dog,
I am no Ox, I am no Hog,
I am not either Sheep or Cow,
Or any beast, that I should bow
To their proud wills, or haughty minds;
Which are as various as the winds:
But I'm a mortal man by birth,
Am born to live upon the earth;
O'er other men I want no sway,
But want my rights as well as they.

---

[43]The calendar, or table of lessons, directed the reader to certain passages in the Bible in the following way: the First Lesson for morning and evening prayer was drawn from the Old Testament while the Second Lesson was drawn from the New Testament. *The Book of Common Prayer* . . . (Baltimore, 1818).

But after all, I do not wish to abcond clandestinely; far, very [far] from my idea, to become a tramp. I wish to live among relations and friends. Will you acquiesce in a plan that I have formed, and give me a little assistance in changing my situation. My plan is this, that you, on the receipt of this letter, procure me some work at Poughkeepsie (any kind of work that I can do, I shall not be exact at first) and then write to my Father, to request him to let me come, without giving him any knowlege, or even insinuiating, that you have had a letter from me. By this means I should leave them in a friendly manner, and they would find me money for the Journey. If you will thus assist me in time of tribulation, you will lay me under the greatest obligations possible; but should you refuse me; then write to me immediately, that I may know to use some other plan. For I am determined to break the bands of opression this once, let what will be the consequence. Whatever way you determin to do, I hope you will make no delay; for I shall wait with unceasing anxiety, till I see the result of this letter. I Still Remain,

Dear Aunt and Uncle;
Your Affectionate Nephew,
Joseph Hollingworth

P.S. We are all in good health. The wedding has not yet taken place, nor does it seem to be any nearer. The Factory will stop as soon as the stock is worked up. Some of the spinners has already done and cleared; all our folks expects to have done this this month. It is said that all Englishmen will have leave this place; work is very scarse in this part, and when a man gets out of employ tis hard for him to get in again. J.H.

[To] Mr. William Rawcliff,
To the Care of Jonathan Hirst,
Wadsworth's Factory,
Poughkeepsie,
Duches-County,
New York.

❧ ❧ ❧

Jabez, George, and Joseph Hollingworth to William Rawcliff

South Leicester January 15th 1830

Dear Uncle

When I left you at Poughkeepsie I had to Pay 75 cents for my fare to New York. Arrived there about 2 O.Clock in the Morning. Left the boat at 7, and got breakfast at No. 7 Gold Street.paid 25 cents. Hartford Steamboats was Stopped, and the fare to Norwich was 6 Dollers in the Fanny[44] & 2 Dollars to Oxford, which was the Cheapest way I could find out. I left New York on Monday at three OClock, and got home on Tuesday Night at 7 OClock. I have not got any steady work yet. Neither do I see any Chance. Our folks is all got into work again. Joseph has never been out of work neither do I suppose he will at this time. I gave them a description of Richmond Factory and they said it was cheap, but they thought it was too

---

44The *Fanny* was in service between New York and Norwich as early as 1824. Fred Erving Dayton, *Steamboat Days* (New York: Frederick A. Stokes, 1928), p. 160.

large for us at present. They seem to be better contented now than before for which Reason I would not advise them to move. As for myself I shall not Tramp any more this winter whether I have work or not and if I should not get a good place between and next spring, I shall come see you again. Mary Hollingworth is 3 feet 6 inches high. William is 2 feet 11 inches high. Benjamin is 3 feet 5 inches high. Hannah is 2 feet 7½ inches high. Your Daughter Mary Ann is 3 feet high. Annas is 2 feet 6 inches high. Hannah weighs 27½ lbs. We all enjoy good health at present, hoping you do the same, so no more at present from your Affectionate Nephew

Jabez Hollingworth

January 18th
1830

South Leicester Jany 17th 1830.

Dear Brother

I embrace this oportunity of writing a few lines to you to apprize you of my advice and opinions respecting certain Matters relative to our futher and final Settlement in this Country. If we intend ever to rescue ourselves from Factory Thraldom we must honestly and sincearly joine in hand and Hart and by our united efforts wisely directed and attended by the Blessing of Heaven we may fairly hope for every reasonable Success. Altho I am an advocate for prompt action yet I would advise to prevent failures that we do not act hastely or inconsiderately but cautiousely and deliberately especialy with respect to our fixing upon a perminent Sittuation for upon this will depend verry much our futher success and prosperity. Son Jabez has told us about Richmond's Factory but I am doubtful we could not at present manage it to certain advantage and I am of opinion that it would be verry hazardous to engague so large a concearn without a sufficent Capital. I think it would be much better when we are able to buy cheap a small Manufacturing Establishment susceptable of emprovment and connected with some land. If this cannot be done that is bought cheap then to try to buy a Farm with a Water privilege thereon sutably situated for Market and then buld or Erect thereon by degrees whatever we wanted always being mindful not to try to fly till our Wings were grown. I am aware there would some difficulties lye in the way but I have not the least doubt but by our united efforts and persiverance that we could surmount them all. I was verry glad when I heard Cousin Joseph Haigh and Family were come to this Country. I shall write to him as soon as I get to know where he is settled. We have been a good while out of work which has been a considerable loss to us but now we are all got to work again. I hope the change here will be for the better. Jabez by staying so long away has lost his place at least for the present. I wish him to try to do somthing for himself. I want him to make a Wagon and if it would not sell I tell him it would transport. I am certain he might make many thing that would hereafter be found useful. Give my respects to all who mirit or are deserving of the appelation of Friends. Fail not to comunicate to us your views Opinions & Intentions.

Yours Affectionately
Geo. Hollingworth

South Leicester, January 17th. 1830.

Respected Uncle,

I understand, by Bro. Jabez, that you are at present a weaver; if so, you will please accept as a New Year's present, the following verses: first written to the few weavers remaining in Factory. South Leicester, December 18th. 1829

Weavers! look round you, and behold
Those empty looms in every part;
Reflect awhile – they will unfold
A useful lesson for the Heart.

Life is a web of tender warp,
Put up within the loom of time;
But Death may come with razor sharp,
And cut it, in its youthful prime.

Then, if the web be not well wove
With filling such as God requires;
The weaver will, with wrath, be drove
Into eternal burning fires.

But, if the work be good and neat,
A glorious premium will be given;
A Crown – a Scepter, and a Seat
Within the Happy Courts of Heaven.

Since this shall be our Just reward,
Let's tend our looms with watchful care;
With filling good, weave our webs hard,
And soon for sudden Death prepare.

J.H.

From your Affectionate Nephew
Joseph Hollingworth,

[To] Mr. William Rawcliffe
at Wadsworth's Factory
Poughkepsie
Duches County
New York

George Hollingworth
to William Rawcliff

South Leicester Jany 24th 1830

Dear Brother

Since writing to you about a Week ago we have changed our minds respecting Richmond's Factory. In that letter I advised to lett it alone at present fearing we were not able at present to manage it. A few days since Jabez mentioned the case to Mr. James Shaw a Saddleworth man who resides here with a large Industerouse Family. This has led to several consultations betwixt us upon the subject and we have finealy agreed and determined to form a Copartnership and engague Richmond's Factory if possoble upon the most reasonable and advantagious Terms. We are aware the greatest difficulty will be to begin or make a start. This over we have not the least doubt of success if we be blessed with Health, for we by our united Families could do all the work and have no wages to pay which is of vital Intrest to the success of a Manufactury. Besides Jas. Shaw is the first man I know who is every way fitted for such an undertaking. He is sober steady Industerous and a good workman, I believe an excellant Carder. He is well informed and has a good knowledge of Men and Things. He has a knowledge or aquaintance with some of the Merchants and Woolstaplers of New York and Albany.[45] This may be of Essential service to our new Establishment. Also he is a man that has a good share of Courage Fortitude and Confidence (viz. what the Yankees call SPUNK) which are necessary requsites for a Trade's-man &c. James Shaw and me has agreed and I hope you will concur, that

[45]The woolstapler was the merchant who bought raw wool from the grower, graded it, and sold it to the manufacturer. *OED.*

you and he and me form a Copartnership on equal Shares and engage Richmond's Factory and use our United efforts to sett it a going, for we can perceive no other mode of extercating ourselves from poverty and thraldom. Besides all the members of our respective Families anxiously Concur. Now what we at present want you to do is emediately on the reciept of this letter to see Mr. Richmond and tell him that we will Engage his Factory provided he will put it into proper and sufficent repairs. We should wish to have it for some length of time at least 5 Years. We are not exact to a month when we enter to it, provided it be not before the 1st. of May. We would rather it was deffered till June or July but you will hear what he says upon these subjects and please to comunicate them to us by letter as soon as possoble. I should like you to get Mr. Richmond to employ Jabez in the repairs because he being upon the Place might not only see what was wanted doing but see that it was done well. If you can succeed in this please to inform us in your letter and we will send Jabez with his tools and every necessary instruction possoble. To prevent a disclosure at this place we shall put this letter into Charlton's Pose Office 3 miles from here, and we request you to direct your letter as follows. —. "Geo: Hollingworth, to be left at Marbles Tavern Charlton, Worcester County, Massachusetts". We at this place are getting into a new Order of things and I have hoped that it might be for the better, but I am not a little afraid that my hopes will be frusterated. They have lowered Weaving to 4 cents per Yd. and it apears to me they intend to have every other thing done as low as possoble. They are posting up a new string of Rules more objectionable than the Old ones. In one of them there is the following "That if any Workman damage any Work or Machinery he shall be liable to pay damage the damage to be assessed by the Superintendant or Agent.

Yours Affectionately
Geo: Hollingworth

Joseph and Jabez Hollingworth to William Rawcliff

South Leicester, March 13th. Saturday, 2 oclock P.M. 1830.

Dear Uncle and Aunt;

Never, before this time has it ever fallen to my lot to write to you on such a Mellancholy occasion; and I am sure that it will be no less shocking to you to hear than it is painful for me to tell that Sister Hannah now lies a Corps! and in a few hours more will be laid low in the silent Grave shut out from our sight for ever! The people are now assembling for to perform the Solemn rites of the Funeral, and I must lay down my pen till it is over, then will I give you the whole of the story.

Sat. 5 oclock, P.M. The Grave now holds the mortal remains of Sister Hannah. She begun her sickness on Wedensday the 3rd. inst. and died yesterday morning about 2 oclock. Her disorder was in her breast, by some called the Croap, others calls it an inflamation of the lungs. I believe it was caused by the Worms, for during her sickness She voided a large Quantity of long round Worms Some measuring about a foot in length. She was Blistered twice on the breast, but it was all to no purpose.[46] It was the will of God that she should quit this troublesome world and not all the Skill, or ingenuity of man could prevent it.

---

[46]Hannah probably died of pneumonia, according to Dr. Henry R. Viets, Consultant for the Historical Collection at the Francis A. Countway Library of Medicine, Harvard University. He says the "worms" were unimportant. "Blistering" was a common treatment to relieve pain by applying a hot iron or an ointment to the breast. Letter to Thomas W. Leavitt, September 21, 1966.

With furious Strides Death moves apace
To terminate man's mortal race;
He spares not, neither Young, nor old,
But with one grasp he turns them Cold;
Their Souls must then to regions go
Of endless happiness, or Woe.

I need not tell you, the consternation that prevades the whole of our Family, occasion'd by this sudden occurance; this, you may in some measure imagine, by calling to mind, that She was in the flower of her age – an only Daughter – an only Sister – beloved by parents, (perhaps the Idol of their Hearts) – beloved by Brothers – the Joy, Delight, and Hope of the whole Family. How true is this Saying, that, "in the midst of Life we are in Death"! Here we behold Sister Hannah in the bloom of youth – sweet as the Rose – charming as Spring – Sportive and inocent as a Lamb; then taken Sick; her days cut short by the Cold hand of Death; and her body commited "to the ground; earth to earth, ashes to ashes, dust to dust," all in the space of about ten days! This is a strong demonstration of the transistory nature of Human Life; and should teach us to "be also ready, for in such an hour as we think not, the Son of Man cometh."

Father and Mother seems to grieve more about this sudden and unexspected event, than any one else of the Family, and perhaps more than they ought to do; but indeed how can it be exspected otherwise? As for myself, though I think a great deal about Sister H. yet I will not murmer nor repine at the allwise dispensation of Porvidence which I believe Is all for the best; perhaps a differant way of thinking aided by my Poetical Muse, enables me to triumph over these difficuties, and make my mind Calm and easy: And in the words of Job to say "The Lord Gave, and the Lord taketh away; blessed be the name of the Lord.'

Come, for awhile let's banish mirth,
Though not afflict ourselves with pains,
For Hannah now lies in the earth –
The Clods doth cover her remains;
    But tho' on Earth her body dies,
    To realms of bliss her spirit flies.

No more her lovly form we view,
No more we see her smiling face
Tinted with Beauty's lovliest hue –
Nor hold her in our glad embrace;
    Cut of by Death, in youthful bloom,
    We have consign'd her to the tomb.

But hark! methinks I hear a song,
By Angels sung – in Heaven above;
There, Hannah Joins the blisful throng,
To sing of our Redeemer's love:
    Around the throne of God they Sing,
    The praises of the immortal King.

Sunday, 14th. Father recieved a letter from Joseph Haigh last Thursday; Joseph H. and family is at Pitsburgh but have not got a situation for work nor do they see any chance of geting work in that parte. It seems that he has been decieved about his Brother Simeon, for he wrote to him from N. Y. and waited for an answer, but recieved none; then he wrote again requesting him to direct an answer to Pitsburgh; but when he got there, he found no letter and when he made inquiries about him, found that "his situation was any thing but enviable." Josph Haigh asks advice from Father for he seems to be in a dillemma. Father is in too much trouble to write him, but he has set Bro. John to do it. I still continue to hate the Factory System as it is here. Give our kind respects to all our friends at Poughkeepsie and take the same to yourselfs. I Remain

Your affectionate Nephew
Joseph Hollingworth

March 14th

Dear Uncle

As Brother Joseph has related to you the loss that has befallen our Family the death of our dear little Sister Hannah I need not say any more on the subject. I shall take up the last mentioned subject. Joseph Haigh requests my Father's advice what is best to do. He says there is plenty of Factorys to sell in the vicinity of Pitsburgh but he says they want as much for them as they cost. But he thinks it is dangerous for him to step into their shoes for if they cannot walk in them he thinks he cannot. My Father is too much troubled to write to him, but he has instructed Brother John to write to him. He also gives him an invitation to meet him at Poughkeepsie, and for them and you to hold a conference on the subject. He also wishes him to write and let us know if he will come and what time he will be there.

With regard to Richmond's Factory it appears to be dropped for very little has been said about it since your letter came, which they did not find any fault with but on the contrary said it was a very sensible one. I have had no work since I left you only what I have done in our own house. I have made a bench to work on which has [cost] me about 6 dollars. I have made a broad loom for Brother John and [. . .] wheelbarrow and some Winterhedges for Sale but I cannot sell them. There is no encouragement for such business here. I am an Englishman amongst Yankees. They want to give me half what they are worth.

March 16 Yesterday morning Father had Notice to Quit as they are going to have all their work done by Girls. Mr. Denny the Agent told Father that they wanted some hands at Southbridge 12 miles from here.[47] Father and I went to see about it but did not make a final agreement. They wished us not to make Application any where else and they would write in a few days. Now you see the Fruits of Large Factorys. Here we are supplanted by Females that is expected to perform the same quantity of work for one half the wages the quality being out of the question. Here we are driven from one Factory to another seeking rest and finding none and when we are in work at what we may call decent wages they have so many different ways to get it all back again that it is impossible to save any thing. The very highest rents fuel Provisions wearing apparrel and every thing else at the very highest prices. The only way to remedy this is unite oursleves I mean our minds and bodily strength together to set about one thing at once and strive to accomplish it. I for my own part has got no money but thank God I am both able and willing to work. I should not be afraid to build houses good enough for any of us to live in.

Yours &c.
Jabez Hollingworth

[To] Mr William Rawcliff
at Wadsworth Factory
Poughkeepsie
Duches County
State of New Yorke

---

[47]Edward Denny, in 1831, leased a mill in Oxford and ran it under the style of the Denny Manufacturing Company, which manufactured broadcloths; between 1835 and 1843 he had an interest in H. A. Pettibone and Company, also of Oxford, which made satinets. Daniels, *Town of Oxford,* pp. 204, 209.

John, Jabez, and James Hollingworth to Tiffany, Sayles and Hitchcock [48]

[April 1830]

We will have the Woodstock Factory and all the Real Estate. Provided we can be furnished with suitable Machinery for Manufacturing of Sattinett Viz. To have the Cards put in to good order with the addition of another Single Machine. A Condencer 12 Power Looms Scouring or Fulling Stocks one Napper 2 Shears 1 Press and its apparatus [. . .] Dying and Scouring Kettles.[49] We will agree to Pay 500 Dollars per Year for 5 Years to commence paying Rent as soon as all the Machinery is ready for Work. Provided you will take the amount of the Rent in Work at a given or Stated price.

---

[48]This letter, those of April 15 and 16, 1830 and the bill of sale dated May 2, 1837 are among the records of the Hamilton Woolen Company, Baker Library, Harvard University.

[49]"Single machine" refers to a carding machine with one large cylinder. The condenser wanted is undoubtedly a Goulding type, since the assumption is made that it will work with enough efficiency to eliminate a slubbing billy. The press was a device used to remove wrinkles from the cloth and give it a shine; it was most probably of the common vertical screw type. The apparatus would include the press papers to put between the folds of the cloth and an oven to heat the iron plates, which were placed at intervals in the stack of folded cloth. David Craik, *The Practical American Millwright and Miller* (Philadelphia, 1870), pp. 399-400.

John, Jabez, and James Hollingworth to Tiffany, Sayles and Hitchcock

South Leicester April 15th 1830

Messrs Tiffany Sayles & Hitchcock

Dear Sirs

We recived yours of the 12th, Inst yesterday, and we have comed to the following Conclusion, Viz. that we have agreed to hire the Woodstock Factory (for the Term of 3 years,) and the Machinary, together with the real Estate, and we will have the Single Carding Machine that is now at Southbridg, and Cards to put the whole in good order, Press and Papers, fulling Stock, 2 Shears, and 8 Looms. We agree to give Five Hundred Dollars per annum, rent to comence first of July 1830. Provided you will agree to the following in addition to the above, Viz that the Looms Shall be Sattenett Loomes delivered at Southbridg, 1 Napper & Dye Kettle that will contain 120 gallons, Delivered at the above mentioned place, Bricks and Lime for to Sett the Kettles, and Press Oven, Shafts gearing and other Materials, for making Drums and puting them in Operation, Materials for making a Tenter and we will find all the work for puting the Same in Operation. By you agreeing to the above Statement we Should consider ourselves Obligated to leave all the beforementioned Machinary in good runing order at the expiration of the Lease i.e. as good as we found it except the natural ware. We Should like to pay the rent in work in Case we can agree on the terms. Our Conditions will be to warrent good work and a price as low can be found elswere. If you dont like to agree to the above We propose to Manufacture Sattenett by the yard at a given Price to have all found and to have no rent to pay. Our price is 13 cents per yard. If you cannot agree to neither of above propositions we must give it up as we dont consider ourselves able to undertake it on any other terms, and it is our intention not to undertake any concern but what we are able to carry into effect.

With respect
J. J. and J. Hollingworth

N.B. Please to write as quick as Possible
Direct as before

⁂

Tiffany, Sayles and Hitchcock
to John, Jabez, and James Hollingworth

Boston April 16. 1830

Messrs. J. J. & J. Hollingworth,

Gent.

Your favor of the 1st. inst. came to hand this morning, proposing terms on which you would like our Woodstock factory.

We feel strongly inclined to accommodate you in case it is not likely to cost us too much, for the money you feel willing to pay us rent for. We have been making some enquiries respecting Kettles, Looms, &c.

There are 8 Looms (New. at Worcester, Ms.) made Stout iron sided for Cotton Shirtings, which we are informed can be altered into Satinett & which will be equally as good as the best Satinett Looms, but the expense of doing the same, we do not know, and were in hopes at the low rent at which we offered you the property, you would be willing to take them at

Worcester, and alter them at your own expense. Some think it will cost 2.50 @ 5$ each Loom, but if you will make the alterations or procure it done, at Worcester & take the Looms there, will allow you Five dollars each loom & pay you the cash for the same, and in addition to our offer of the 12. inst. will procure a 120 Gall. Kettle & send the same to Southbridge, and as respects a Napper if we correctly understood you, it is merely a cylinder clothed with cards, same as the clothing of the carding machines, & if so you can find enough of these about the factory at Southbridge & a plenty of the second hand clothing which you shall have for the same.

Also Bricks and Lime to set the Kettles; and build the press oven, Shafts & Lumber for making Drums for said machinery, and also, I presume we have plenty of the press paper such as we formerly used for our Broad Cloths, which we will spare you. The Rent to be five hundred dollars paid in cash quarterly, unless we should subsequently agree with you to work up our coarse wool, & waste, which is most probable in case you make good goods, & at fair prices. The 8 Looms we refer to, are at Samuel Clark's factory New-Worcester on the Oxford Road; and you can find a good Machinist close by the factory, who is in the way of making Sattinett Looms, & can alter them cheaper than we could at Southbridge.[50] Should be pleased to hear from

[50]At least two firms were manufacturing looms in Worcester at this time. Barnes Riznik, "New England Wool-Carding and Finishing Mills, 1790-1840" (Sturbridge, Mass., 1964), pp. 59, 62; U.S., Treasury Department, *Documents Relative to Manufactures,* pp. 572-573. In order to adapt all-cotton looms for the manufacture of satinets, the shuttle probably had to be adjusted in order to accommodate the wool yarn. Samuel Clark's "factory" apparently grew out of a fulling mill, which he operated for some time on Kettle Brook in Stoneville, a village in that part of Worcester which became part of Auburn in 1837. *History of Worcester County* 1: 245-246, 248.

you soon as convenient as we intend to advertise the property for sale in case we do not let it to you.

Respt. yr._____________. Tiffany Sayles & Hitchcock

[To] Mr James Hollingworth
South Leicester, Ms.

❧ ❧ ❧

Joseph Hollingworth
to William Rawcliff

Globe Village. Southbridg, Mass. April 18th. 1830.

Dear Uncle and Aunt,

"Hoot away Despair,
Never yield to Sorrow;
The blakest sky may wear
A sunny face tomorrow."

I now take up my pen to inform you, that Bro, James, (after much searching,) has at last found his Fair Rib; or in other words, that he has got M-a-r-r-i-e-d.

Now Jemmy has resolv'd at last
To take to him a Wife;
The ceremony it is past,
The contract is for Life.

But yet one thing remains for Jim;
That he must treat his Molly,
As he would wish her to treat him
Or marriage is all folly.

Now B-t-y B——k may grieve and sigh,
When with a retrospective eye,
She views the days that are gone by;
When Jim and her, by the coal house door,
Or else perhaps on Nether Thong moor,
Enjoy'd themselves, in days of yore.

The following is a coppy of a letter, which I have caried to the Southbridg printing office, to be published in the Register.

"Married, on the 11th inst. at Wilkinson's Ville, Sutton, by the Rev, Mr Goodwin, Mr James Hollingworth of Southbridg, to Miss Mary Shaw of South Leicester.[51]

Cupid is busy with his darts,
And so is Hymen with his bands;
The former strikes the youthful Hearts,
The latter gladly Joins their Hands." ————

Massachusetts

To Wilkinson's ville, as I'm inform'd,
Three dashing Couples took a ride,
Not careing how the weather Storm'd,
Determin'd to have the knot tied
There was Ann Shaw, with her Brother Jo,
And Jabez with his spark
Likewise the Bride, with the Groom by her side,
All sprightly as the Lark
But I, the meanest in creation
Did never have an invitation!

---

[51]Rev. Goodwin was pastor of St. John's Protestant Episcopal Church in Wilkinsonville. *History of Worcester County* 2: 381.

The following lines, first written to Counsin James Hollingworth after his marriage in 1829, will be applicable to Bro. James, at present

What! dost thou think I am to vain?
Or dost thou think to give me pain?
Or what else made thee disdain?
    To ask me to thy Wedding O.

Once, I did, thy name revere,
But now of late thou hast grown queer,
Thou wast afraid I should appear
    At thy Almighty Wedding O.

An ignorant Clown some thinks I be,
Not fit to appear in company;
And so I wasn't thought fit by thee
    To come unto thy Wedding O.

Believe me Sir, I do not care;
I might have been caught in some snare
Had I, been summon'd to appear
    At thy Almighty Wedding O.

The time, I guess, may soon arive,
If God permit that I should thrive,
When to myself I'll take a wife;
    But thou shant come to my Wedding O.

May 23rd. After I had written so much of the letter I read it to the Bride and Bridegroom, they was vexed with it; said that after I had written such a letter, as the last one I sent to you, I ought to be ashamed of writing such stuff as this. But the main thing is, I have told that, I was not invited to the wedin; but it is truth, and why should I be ashamed of telling the truth? The reason that they alledge for not inviting me, is somthing like this, THAT THEY WAS AFFRAID TO EXECITE OLD SHAW'S JEALOUSEY, AND THEREBY LOSE HIS FAVOUR! But have they not forfeited more HONOUR, than they will ever gain favour? Father lately recieved a letter from Joseph Haigh at Pitsburgh; neither him, nor family has yet got any steady employment. He has been on a tour into the State of Ohio. He gives a good acount of the country. Nay I think tis charming!

Brothers John & Jabez & Cousin James Hollingworth has hired the Muddy-brook-pond Factory in Woodstock, Connecticut. on a lease of three years. It is about 5 miles from this place & about 100 rods from the line devideing Massachusetts from Connecticut. The place is owned by the Same Bostonian Company that owns this place.[52] The Factory is supplied with watter by a natural Pond or Lake which covers an area of about 100 acers. They intend to Manufacture Sattinet & do Custom work. We left South Leicester & came to this place about the 1st of April. I believe that the reason we had to move, was the effects produced by Masonic power and Yankeeism;

[52]As early as 1820 there were three woolen factories in Southbridge, with the following machinery: eight carding engines, four slubbing billies, four jennies, twenty-seven power looms, two shearing frames, three shearing machines, two pickers, and two gig mills. The James Walcott Woolen Manufacturing Company (Plate IV), successor to the Globe Manufacturing Company (1814), made broadcloths and cassimeres between 1820 and 1830 when the owner sold the mill to his selling agents, Tiffany, Sayles and Hitchcock, 82 Kilby Street, Boston. "Questions to . . . Persons . . . in Manufacturing Establishments . . ."; "Brief Record of the History of the Hamilton Woolen Company . . . ," filed with the company's records, Baker Library, Harvard University; *History of Worcester County* 2: 301; William R. Bagnall, *The Textile Industries of the United States* (Cambridge, Mass., 1893), p. 567.

Plate IV

Globe Village, Southbridge, Mass., 1822. Oil, by Francis Alexander. Courtesy of the Jacob Edwards Memorial Library, Southbridge, Mass.

when we quit I left the following lines within a few rods of the Taveren, in South Leicester.[53]

1st. Come all ye Anti-masons, where ever you do dwell,
Come, unite all your efforts to break the Mason's spell,
Destroy all their Lodges, their "mystic knot" untie,
Then sink them in oblivion, and there let them lie.

2nd. Rise up ye Anti-masons, and all united be,
Against this Hell-born Monster, I mean free-masonry;
O spurn all its precepts – its principles despise,
For 'tis all Imposition, Hypocrisy, and Lies.

3rd. Hark 'tis the blood of Morgan, that calls you to the field,
To fight the Monster boldly, untill you make yield;
And when you have Subdu'd him, which certainly must be,
O then erect the Standard of Truth and Liberty.

The name of South Leicester is changed to Clappville.

This place and the sorounding Country is a great deal more Pleasant than So. Lestr, But I cannot say that I like the place any better. Father is warping on a masheen, Edwin is Spooling, James is Spining on a Jack, Josph Kenyon is Roping & Mary

---

[53]In this, as in other matters, the Hollingworths demonstrated their ability to be as American as the natives. "Morgan" was of course William Morgan (1774?-1826?), the Mason whose disappearance just as he was about to publish a book on Freemasonry gave rise to rumors he had been murdered. *DAB*, vol. 7.

weaving.[54] Joseph boards at our house, but Mary does not. I have not had any steady worke here yet nor do I expect to have; we are all in good health at present hoping you and yours are the same. I now am constrained to coincide with the general opinion, that crossing the Atlantic does really change men's affections, for while you remained in England we kept up a regular corespondence, but now that you have crossed the Atlantic it is all over, but what is the reason? Do I write to often and thereby cause an unnessary expenc of postage? or when I write dont I chuse propper Subjects and use propper language? I hope that you will write on the reciept of this and if I have commited any offence tell me plainly, and I will endeavour to make amends. Direct for Joseph Hollingworth Globe Village, Southbridge, Mass. and you will oblige your

Still loving Nephew
Joseph Hollingworth

Though Cold is the Grave where Hannah doth Sleep,
And distant the Spot – yet I will not weep;
And though I dont fret – yet I will not forget,
The prattle and smile, that did sorrow beguile,
But Dear in my mem'ry the Relic will keep.

[To] Mr William Rawcliff
at Wadsworth's Factory
Poughkeepsie
Duches County
New York.

---

[54] "Roping" would have involved tending the condenser at the end of the carding machine and transferring the spools of roving (or roping) to the spinning jack.

Bradley Clay
to William Rawcliff

Huddersfield 5 June, 1830

Your letter of the 2nd ult. arrived on the 2nd of June and your letter of Attorney came the following morning. I have only received one letter for you, since you left England. It is from Joseph Hollingworth dated the 12th of July, 1829. The following account contains the items I have received and paid since you left: namely

Dr. ———————William Rawcliffe———————Contra———————— Cr.

1829			1829		
Aug. 26	to Star__(?) Club	4. 7. 9	Oct. 27	by Thomas Crooks	3. 2. –
Nov. 25	———Do———	4. 7. 9	Nov. 24	———Do———	5. –. –
1830			1830		
Feb. 24	Do	4. 7. 9	Feb. 23	Do	5. –. –
May 26	Do	4. 7. 9	May 25	Do	5. –. –
		19.11–			19.11–

To Balance owe to me 1. 9. –

I have your cart in my possession yet. Mr. Crowther could not meet with a customer for it, nor have I been able to sell it, or even get any thing bit for it. I note what you say about different individuals owing you money and will endeavour to collect them in immediately. Jonathan Senior is dead, but his son who has succeeded him will probably know something about the stones you mention, and I will see him next Tuesday on the subject. If I should have occasion for an Attorney in your business, I will attend to your recommendation. I do not know when I shall meet with a customer for your estate but I will see Mr. Wilson of Netherthong and Thomas Crooks about it; perhaps I may purchase it of you myself for money is not worth more than three per cent in England.

All the Manufacturers in this neighbourhood are now very busy, and every thing is getting cheap. Beef and mutton is now /5d per lb. Fresh butter /10d and some think it will be /6d as the summer advances. New milk /1½d per quart. In short when our Government reduce their prices, which they must do soon, in proportion to other things, this country will be one of the cheapest, as it is the best to live in, in the world. We are all becoming Reformers, that is to say, the labourers have been so some time, and the middle classes are fast joining them. I do not now despair to see this Island become the best country in which a man with moderate wishes, would choose to live in. I wish to caution you against one thing, which is "Spirituous Liquors." It is the bane of the country which you have chosen to adopt. Avoid the glass, as you would slow poison.

You may rely on my best attention to your interests in your native land.

I am your friend,
Bradley Clay

To Mr. William Rawcliffe
Poughkeepsie,
New York.
U.S.

John Hollingworth to William Rawcliff

Woodstock July 4th 1830

Dear Uncle

I write to inform you that we are all in good health at present and hope these lines will find you the same. We have made a genral move this Spring. My Father and family are at Southbridg about the same Distance from here as South Leicester is from Oxford. My Father and James, Joseph, and Edwin are all in work at the above mentioned place. James has got married about 2 Months since. Brother Jabez cousin James and I ha[. . .] at this place to manufacture Sattenette. We came here about the 1st of May and our rent comenced on the 1st ints. We are to pay 500 Dollars per Annum for 3 years. We have Seven houses. The Factory consists of 2 Building connected together 3 storys each. They are about 18 feet wide and 36 or 40 feet long. We have 3 Double Carding Machines and 1 Billy 1 Jenny [a] Picker and Fulling Stock 2 Shearing Frames 1 Press and 1 Dye Kettle.[55] We are to have 6 Power Looms which is to be ready By 10 inst. There is about 2 Acres of Land a Pond of about 100 or 150 Acres which we can draw down 10 feet. It is quiet a pleasant place in fact it realy would be desireable if it was sittuated within 2 Miles of the North River. I forgot to say that there is 15 hand Looms at the place which we can use if we want. There is a Barn and other outbuildings.

---

[55]A double carding machine is one with two large cylinders in the same frame. A picker is a machine for untangling the wool fibers before carding. The shearing frames may have been the special type of bench used with hand shears, or they may have been a mechanized form of hand shears, which preceded the rotary shearing machine.

But to Change the Subject I must inform you that we have had 3 Letters from Joseph Haigh and his son Uriah. 2 from Joseph was directed to my Father and one from Uriah to Jabez. My Father recieved his first about the time that my Sister Hannah Died so he desired me to answer the letter for him. He wished me to send him an extract from one of your letters about your Journy to West of what you saw and the information you received about that part of the country. Since then we have received the other 2 directed as above stated of which I shall send you some extracts from them containing some Questions there present Sittuation and State of Mind together with the Price of Provisions, which I call ruinous Prices. I Shall now give you the extracts which are as follows, "Your Son John mentions a place that lies between the Hudson and Delaware Rivers which your Brother William has seen and as I understand he recommends very much. He mentions a Mr Parks who owns a Grist Mill Saw Mill, and 2 Turning Mills 2 houses with 67 Acres of Land.[56] I do not know what he means by the Turning mills wether they Be 2 Seperate Building or only one and I could wish to know wether any of the Mills could with a small expence be converted to our Business or not. He sayes he wants 3000 Dollars for them but he does not mention his Terms wether all [. . .] money and Credit for [the] remainder for a given time. He does not mention wether all the Land be under Cultivation or what Proportion of it. These are Questions which I could like answering. I could like also to know what the

[56]A turning mill was an establishment for performing work on a lathe. In 1820 it could have been water- or hand-powered. Edward Hazen, *The Panorama of Professions and Trades* (Philadelphia, 1837), pp. 219-220. The property he is describing was owned by William Parks of Liberty, N.Y. Parks was something of a titan in Liberty; he and his son, Elijah, built several mills there on the Little Beaverkill River, and the village of Parksville was named for him. Quinlan, *Sullivan County*, pp. 338-339.

Quality the soil is of what Quantity of Corn or Wheat can be raised from an Acre and also the prices of the various kinds of Produce at home or in the Market, and wether you think the place where your Brother William speaks of will be the place where you intend to make a Settlement or wether you think it would not be adviseable to make a settlement in the State of Ohio. I am quite Ignorant of the Situation that he mentions. Just mention wether there be any Coal in that part of the Country and also if there be a good many Farmers in the Neighbourhood." The above is from J. H. Letter. I shall proceed to give you Uriah's, "I wish that my Father had gone to meet your Father at Poughkeepsie when he came back from Ohio. The reasons that he did not were first that he had been at a great expence in traveling through the State of Ohio, Secondly he expected some letters from that State but he has been disapointed for he has not received any Thirdly he wanted to know more particular about the Sittuation that your Uncle William Speakes of. Now I hope you will write imeadiatly and as far as possiable answer the questions in my Father's letter. There is a person who is now living in the next house to us, but who lately came from Newburgh. He sayes that he now wishes that he never had comed here for the farther he came west the farther he came astray. He sayes also that he has seen the Cannal that connects the Hudson and the Delaware Rivers and he sayes that the Country around is a fine and healthy Country. We are compleatly tired of Stoping here doing Nothing." U.H.

I will now give a few more lines from Joseph's Letter as follows, "I had better have given 150 £ than have comed here. We shall wait with Impatience your answer. Give us all the information you can and if you think it would be adviseable we will com all at once for it is of no use stoping here doing nothing. I shall enumerate a few of the prices of Provisions,

Superfine Flour \$2.95cts per Bbl. Beef from 3 to 5 cts per lb Dried Hams 8 cts per lb Shoulders and Fliches 6 cts p[. . .] per bushel, Besure you do not neglect one moment to write. I am Sirs Sincerly

Joseph.Haigh."

We wish you answer the above Questions and write directly to Joseph Haigh to Save time as his last letter has been 3 Weeks in South Leicester Post office and he wanted an answer imeadatly. Therfore we hope you will not fail to write to him without Delay. Direct for him at Pittsburgh Post Office till [. . .] called for. By so doing you will greatly Oblige Us all.

I shall now give you our opinion which we think will be best plan to act upon in the present Case. As we have hired this place and as we have more room in the Factory than we can occupy ourselves and more houses than we shall want we think it would be best for them to come to this place and help us to occupy it for the term that we have hired it for and in the meantime we could be preparing another place or buy this which we think the most proper. Give our respects to my Aunt Nancy and to rest of friends at Poughkeesie and take the same to yourself. I should like you to write to me and let me know how you are coming on. When you write to any of us Direct to Southbridg Worcestr County Mass. I am with respect your

Aff Nephew
John Hollingworth

[To] Mr William Rawcliff
Poughkeepsie
Duchess County
St of New York

Joseph Haigh to William Rawcliff

Pittsburgh 8th July 1830

Mr William Rawcliff.

Sir,

I wrote your Brother in law George Hollingworth on the 4 of March giving him an Account of our disapointment with coming over the Alleghany Mountains and wishing him to have the goodness to let us know if he knew of any place that he thought would suite us, and received for answer that he did not know of any place in their Neighbourhood. The Letter was Wrote by his Son John and he wished me to write immediatly and fix a time when I would meet his Father at Poughkeepsee where He — I — and you should have a personal conferance together to propose and adopt such plans as would tend to the benefit of us all. He says it is our intention to settle in one Neighbourhood and we should like to have you for a Neighbour. He also says that you had been with your Brothers in Law 50 or 60 Miles West of Poughkeepsee to look at some land and that they had purchased 159 Acres part cleared. He says that you like that part of the Country very well being well supplied with clear, soft Water and the Country said to be Healthy. He says Land is Cheap and being handy to Philidelphia or New York for a Market, perticulerly the latter, as there is a Canal runs thro the tract of Land that lies betwixt the Delaware and Hudson, River's. He further says that you Mention a Mr Parks who owns 3 Mills with 67 Acres of Land belonging to the same, and that he wants 3000 Dollars for them. When I received his Letter I had made up my Mind to have a fair view of the Ohio State and we having to move from the place that I had rented in Pittsburgh on the first of April and having another place to find I could not get of till the 2th therefore I defered Writing till I returned. I wrote him an answer on the 28th of April giving him a short Account of my Journey and have waited for further communication until I thought I had given a sufficient time for an answer from him. I order'd my Son Uriah to write to Jabez and he wrote on the 5th of June and have been waiting for an Answer but no answer have we got from either of them. I wrote a Letter to Willm. Haigh who lives in Hammond Street N°. 48 New York on the 18th of June and received an Answer Yesterday requesting him to say whether he knew of a situation that he thought would suite for our purpose and also wished him to write if he knew any thing about our Cousin Hollingworth or you. He says he beleives, but is not certain that they are removed from South Leicester and that you are at Poughkeepsee, and he was informed it was your and others intention as soon as you could see your way clear, to make a settlement somewhere, near the Delaware River, but perhaps it would be sometime before you would put it in operation. I am very much surpris'd that we have received no Account from our Cousen Hollingworths. We feel very much disapointed – as it is high time that we were settled somewhere. Now Sir what I request of you is and not delay one Moment Writing an answer as we are upon the point of Making a Settlement in this State or in the Ohio, and could wish to know your and Hollingworths intentions before we do make a Settlement. I could wish your Opinion respecting M^{r}. Parks Mills whether they have water enough in the Summer Season, also whether any of the Mills could at a small expence be converted into our buisness. I could like as I cannot get my Family Work for Wages to get a small Carding Fulling and Finishing establishment, to be employed by the Country Farmers, as I understand it is the custom in America to

Manufacture a part at least of there own wool for there Domestic uses.[57] Please to say whether you think that the beforemention'd situation would be a suitable one or not, and also say whether the Land be under cultivation, if all or only a part, and what proportion and any other information that you think will be of service will be thankfully received by yours,

truly — Josh. Haigh

P.S. As Cousen Hollingworth does not say what are the terms of Mr. Parks whether all Money down or a part Money and credit for a given time for the payment of the remainder, if you know the terms be so kind as let me know what they are. We are all in pretty good Heath at present — hoping you and yours are the same. Give my respects to Willm. Lockwood and ask him if he could not like to go to England again. Let us know how Mr. Wadsworth is coming on with his Factory &c.

[To] Mr. William Rawcliff.
Poughkeepsee Dutchess County.
New York State.

with Speed

---

[57]The carding mill was part of the American landscape long after the factory system had arrived. Old Sturbridge Village, Massachusetts, found one intact and equipped in 1955, in South Waterford, Maine. It has since been moved to the Village, where it has been restored to operating condition. The Merrimack Valley Textile Museum has in its collection several carding engines acquired from similar mills in West Virginia.

Joseph Hollingworth to William Rawcliff

Muddy Brook Pond Factory,
Woodstock, Connecticut Sept. 5th 1830

D—evolve on me the pleasant task,
E—ach time, to answer what you ask;
A—nd in return for favours done,
R—elate how things are going on.

A—ssisted by a power devine,
U—nveiled before me truth shall shine:
N—ature's grand works may all decay;
T—ruth shall endure to endless day.

A—nd now I take my pen in Hand,
N—ot doubting but you'll understand;
D—esiring, that you wont mistake,
U—nknown, the errors I may make.

N—ow tho' to distant lands we've roved,
C—an we forget those whome Loved;
L—ove, the great source of all our weal,
E—nlightens every mind with Zeal.

I recieved yours of June 20th. on the 5th. of July. which gave me great sattisfaction. The reason that I did not write sooner is this, Bro. John had written to you Just before I got yours giving you a description of this place, and also desiring you to write to Joseph Haigh, so I concluded not to write then lest you should have letters coming in to thick. Bro. James and Wife lives with Father and family at Southbridge. Mother says she should like to have come over to see you but that she could not well be spared. You want to know what work I have. Well, when I came to Southbrigdde at first there was no work for me at my old buisness, so Mr Sayles set me to work with my Father at the warping macheen.[58] I worked 5 weeks when I thought it time to ask what wages I should have. The reply was NOTHING! that having the chance to learn a fresh trade was thought a Just compensation for my verry valuable services. The result of which was, that I got into a Jackass' fit. Father then took the warping and spooling by the Job. He and Edwin worked at spooling and I at warping untill I got weary of the work. I then came here to work, when M^r^. Sayles sent for me back, as he wished to hire me to work in the fulling Room for a few days. I went back and worked 21 days for 12 dollars. And finaly I came here again, and am going to do the Napping, Shearing and pressing, when the work is ready. Mary Kenyon has had the misfortun to lose the forefinger of the right Hand. She was weaving on a power loom. She put her finger where it had no buisness, and so the loom in return snapped it of between the first and second Joints.

That sentence where you talk of 8 Brooks meeting at W. F. is a verry good pun, if I rightly understand it.

And now for a description of this place. It is situated in the township of Woodstock in the "land of steady habits" alias Con^ct^. about 4 miles south of Southbridge and about 16 from South Leicester alias Clapville. This place contains about 3 acers more or less on which is the Factory, consisting of two buildings Joined together, each 3 stories high, a dye house, a wood shed a Barn, and seven houses or tenements, together with another building divided into 4 sheds with a large chamber over,

---

58Willard Sayles, of Tiffany, Sayles and Hitchcock. "Hamilton Woolen Company . . . ."

which may be used as dry house. In the Factory are 3 carding machiens, 2 billys, 4 Jennys, 13 broad hand looms, 4 new satinet power looms, 1 fulling stock called a poacher, 1 picker, 2 broad shears, 1 press, 1 dye kettle, 1 satinet Napper and 2 shears which we have had to buy, and several other things.[59] There is a most excelent watter weel, an over shot weel the best I ever saw.[60] It is suplied with watter from a larg pond called the Muddy Brook pond, although the watter is as clear and as soft as any other. They have hired this place for $500 a year, for 3 years but will have to Quit any time the owners think fit at 12 months notice. The owners are now determined to sell it without delay. They ask 6000 $ and will not take less. There is a party of Yankees wants to buy, but they say they will give our folks the first chance and make the payment easy. Joseph Haig, Father, Brothers John, Jabez, & Jame and Cousin James have determined to buy rather than quit the place. The interest of the money will not be so much as the rent. Joseph Haigh & family arived here last Wedensday and I believe are verry much pleased with the place.

If you could make it convenient to come over, and see us, and the place I should be very glad. Perhaps you would Join our folks in buying of it, or if not sold you might like to buy it youre self. It might be a good place to keep a store, the nearest being 3 miles of. If you come you may come by way of N.Y. and Hartford, from thence by the stage to West Woodstock a place 4 miles from here, and I should like to see Aunt Nancy come along with you.

---

[59]"Poacher" was apparently taken from the verb "to poach," one contemporary meaning of which was "to mix with water and reduce to a uniform consistency," according to the *OED*.

[60]The over-shot water wheel was one which was filled from the top: the force and weight of the water entering the buckets moved the wheel that turned the drive shaft. Evans, *Young Mill-Wright*, Plates 13-16.

Bradley Clay
to William Rawcliff

Huddersfield 7 September, 1830

Sir,

Ely Hobson of Thong offers two hundred and sixty pounds, and Thomas Crooks two hundred and fifty pounds for your estate at Miry Lane Bottom. Now it is possible that I might get from ten to thirty pounds more from one or other of these persons, and that I believe is the full amount you will get for it. May I request your answer with your opinion, as to what I ought to do.

Trade is now very good, both woollen and cotton, every body is fully employed and wages are getting up.

All our factories have orders in advance for the next six months, and I never recollect the Cloth Hall so clear of goods as at present. Cloth is full twenty five per cent dearer than it was six months ago. Let me hear from you soon.

Yours respectfully,
Bradley Clay

To Mr. Wm. Rawcliffe,
Poughkeepsie, N.A.

Joseph Hollingworth to William Rawcliff

Globe Village, Southbridg, Mass. Nov^{br} 2nd 1830

Dear Uncle

I recieved yours of the 21st of 9th month in due time. Have delayed writing till now, can not apoligize for so doing, only that Procrastination is the theife of time. You will be apt to think that I am a rambling sort of Chap, when you see that I have got to Southbridg again, but you shall know the reason why and wherefore. Father has had the misfortune to burn his arm so badly that he could not work, and so I have come over to do his work till he gets well again. The Girls are all very Glad to see me back again. You may guess the cause as well as myself, but Alass! how greatly they will be disapointed. I think by your manner of writing that you grow verry Quakerish. LORD KENYON has quit boarding at Fathers, and is gone to board with Squirer Plimtons.[61] The Poor fellow is too Rich to pay for his Board. He has left 27 dollars due.

Big with importance, now he struts,
And tries to hold his Noddle higher:
He thinks it is a famous thing,
To be a Boarder with a Squier.

The Woodstock Factory is not sold yet, and probably wont be. I stated in my last letter, that Father, Brothers J, J & J, Cou,

[61]"Lord Kenyon" is Joseph Kenyon, who went on to become a mill owner, perhaps with the money he saved by not paying his board. See fn. 71. "Squirer Plimton" was Henry Plimpton, who kept the tavern in Globe Village. Levi B. Chase, *Plimpton Family Genealogy* (Hartford, Conn., n.d.).

J, & J. Haigh had agreed to buy the concern, so I was informed, but I have scince learned that such an agreement never existed. I have finished about 500.^{yds} of satinet and have made out better than I expected to do. Mother would have been glad to come over, but she could not be spared. Father says that if you take a Factory, she may come live with you, and so will he. If you do take a Factory I also would be verry glad to come and willing too.

Muddy-Brook-Pond Factory, Woodstock Con. Nov 8th 1830

Since Writing the foregoing I have learnt that Father & Joseph Haigh intends to come see you in the course of a few weeks, but have not heard the exact time stated. I should have told you somthing more about our Factory concerns but think it unnesessary as Father is coming. I am Glad you did not come, as I requested in my last, because I now see with different eyes. I like the Place itself, as well as I ever did, but the concern is too much in Yankee fingers for me.

Yankee doodle dandy,
The Yankeys they are handy,
To rogue and cheat,
And make folks sweat —
To smoke Segars, – and drink a glass of Brandy.

I have been informed that Jemmy Anderton, (alias "old Buckram",) South Leicester, (now Clappville) old superintendant, and Slavedriver, is, together with his concubine (Fanny Wilby) residing at a Factory somewhere between Albany and Buffaloe. He is the Superintendant or Slave driver and recieves

5 $ per day. I also understand that you have got Gorge Mellor's Note of 20 £ against Anderton. If so, I would have you look Sharp, and catch the old Mason. I came back here on thursday last, and have finished 200 yds. more of Satinet. If you do take a Factory I should think you will have work for me and I will come any time when you think fit. But dont you think farming the best, and surest way of getting a living? Manufacturing is a very unsteady buisness, somtimes up, and somtimes down, some few gets Rich, and thousands are ruined by it. Rogues, Rascals, Knaves and vagabonds are connected with it. Some persons that you trade with will cheat you in spite of your teeth, and you must cheat others in return to make ends meet and tie. In Short no honnest man can live by it. A Factory too, is liable to be burnt down, but a Farm cannot be easily burnt up. Manufactoring breeds lords and Aristocrats, Poor men and slaves. But the Farmer the American farmer, he, and he alone can be independent, he can be industrious, Healthy and Happy. I am for Agriculture. I am young Just steping into the world. I may probably be married somtime, and have a family, but I cannot bear the idea, that I, or my children (if I should ever have any) should be shut up 16 or 18 hours every day all our life time like Slaves and that too for a bare subsistence! No, God forbid. If I had the chance to morrow of either a Factory worth 10000 dollars, or a farm worth 5000 dollars, I would take the Farm. But after all, I would say please yourself, you are older than I, and knows the world better. A small Factory with a quantum suficit of land along with it might do pretty well if well managed. I believe we are all in good health at present hoping you are the same. Write as soon as posible so that I can answer you back again when Father comes. Is Aunt Nancy got better again? How is Mary Ann? And the other one, Anis I beleive her name is? Is your youngest, a boy or a girl? How is your mother,

and sister Hannah? And where is Johnathan? Send me the price of Jamisons Atlass, and you will oblige your aftionate Nephew

Joseph Hollingworth.

[To] Mr William Rawcliff,
Wadsworth's Factory,
Poughkeepsie,
Duches County,
New York.

❧ ❧ ❧

George, Jabez, and Joseph Hollingworth to William Rawcliff

Pond Factory, Woodstock, Con. Feb.y 27. 1831. Sunday

Dear Uncle,

I recd yours, of the 17 inst. last night, and I take this early oppertunity of writing again. It appears to me by your manner of writing that you are uneasy, and unsettled, but cheer up, be not discouraged for great is the land that is before us. Before I begin to answer yours I must write somthing to make you laugh, for the Proverb says "laugh and be fat". Very well, Mary Hollingworth got to bed of a great lad about the middle of December last.[62] This occurence drew forth from the Poet's

[62]Frederick Monroe Hollingworth, the first of James's children, was born on December 16, 1830 in Southbridge. He was followed by Washington Irving, May 5, 1832; Benjamin Franklin, December 3, 1834; James Byron, April 27, 1836; David Milton, January 20, 1838; and Hannah Frances, March 30, 1842. Southbridge, Mass., *Vital Records.*

knowledge Box somthing in the following strain,

> Let every heart be full of Joy
> To banish Hyps and dull spleen
> For Bro. James has got a boy
> Such as I guess was hardly seen.
> It is as big, I snum it is
> As many a two year older;
> And then his great big looking Phiz –
> Who ever saw a bolder?
> But then the strangest thing I say,
> Which seems to me both strange and queer
> 'Twas but 8 months from the wedding day
> When the young baby did appear.

The reason that Father & Jos. Haigh did not come as I said in my last, was because Father was busy and could not leave work without danger of losing it. We have some thoughts of settleing at this place. It is proposed that we, viz Father, Bro. John, Jabez, James and James Hollingworth & my selfe, take this place at Six thousand dollars on these terms that from each and every yard of satinet made in the Factory five cents per yard shall go for the payment of the place till paid for, that the owners shall give us a deed of the place and we give them a mortgage on the same. We have ordered 4 new looms which will be ready soon. Our market is Boston. We have engaged a Commission Merchant to buy our stock and sell our goods. Joseph Haigh does not join us. He and Family are at this place, but I understand they are looking out for another, wether they will go or stay I can not tell as yet. Father and James are still at Southbridge. They are come over here to day. I have showed them your letter, and John & Jabez also, and they all are very anxious that you should come over here and see the place and us. You seem to be out of work and unsettled. I think if you was here we could find you work. However you might do well to come and see, perhaps you might like to Join us. At any rate there would be no hurt to come and see for yourself, and I could like you to come imediately, if you and Family be in good health. Then we could discus over things verbaly, face to face better than by letter. I therefore repeat, desire, and intreat, that if you can without any inconvenience to yourself, to come and see. Mother has been sick, but has again got better. Not having much more to write at present

I subcribe myself   Your Affectionate Nephew   In good spirits,
Joseph Hollingworth

P.S. If you come please to Bring the book along with you and I will endeavour to make recompence.   J.H.

[P.S.]
Dear Uncle

As we shall expect you here to see us in a short time, and as I have not much time to write at present I shall only say a few words. On the 17th Jany I was taken sick of the ear ache and was confined to my room for about 3 weeks. I am now better but I am a little deaf. John has some little sickness from Colds. If you should come as I hope you will, I should like if you can conveintly bring my spirit level with you. But I do not wish you to do it unless you can without much Trouble. If I did not expect to see you soon should Try to say more. So I remain your Afectionate Nephew

Jabez Hollingworth

[P.S.]

Respected Brother

I also invite you to come over and see us if you possably can, as I have a great deal to say to you respecting ours and your futher prospects. We have made a Bargan for the Pond Factory and are to make the Writing deeds this week but I think I shall endeavour to put this off till you come here. I am of opinion this is a favourable oppertunity for us [. . .] we sett to it in good earnest. I have some idea if you were to fall in with us in the concearn and you to situate yourself in New York as a kinde of a Comission Merchant you might be useful to yourself and to us. But more of this if you should come which do as quick as possable,

I remain Yours &c
Geo Hollingworth

Globe Villiage
Feby 28th 1831

[To] Mr William Rawcliff
Wadsworth's Factory
Poughkeepsie
Duches County
New York

Joseph Hollingworth
to William Rawcliff

Pond Factory, Woodstock, Con. Sunday July 17th, 1831

Uncle William

I received your last letter, dated June 19th on the 3rd inst. by the hand of Bro. James. It came to me with the seal broken open! James told me that Father had paid for the letter at the post office and therefore thought himselfe entitled to break it open! Disgraceful! Shameful!! (and if there be any sin in the world) tis Sinful!!!

I confess I feel a little ashamed to think that you have wrote letters without my writing any. But I waited day after day, week after week, and month after month; so that I might have Somthing to write about, but all in vain; and now sit down to write a letter about nothing.

Sunday 24th. Thus far I wrote last Sunday when I was interrupted by a large congregation of Hypocrites who assembled at our house for the purpose of holding a Religious meeting as they are pleased to call it. In your letter of the 13th of March you say somthing about going to your land in Sulivan County [N.Y.] and building a log hut; I should not advise you to do that because I think that you could not be so comfortable as on a farm with good Buildings. In my last letter I said that I was one of the Company here; It was so proposed, and at first I accepted the proposal; but when I thought more deliberately about it, I again declined. And this I did for several reasons. — 1st Because I did not wish to becom rich suddenly. 2nd Because I did not want to get into debt. I am poor enough already, but then I owe nobody anything. (excepting you; but I dont know how much.) 3rd When I settle down on a place I should want to have some land, at any rate enough to make sure of a living. I mentioned in my last that Joseph Haigh was looking out for another place. He did so, and found one at Milbury,[63] a place about 24 miles from here. He and his Son John went to work a month on trial. The rest of the family staid here. They staid about 2 months, but not likeing the Confinment, Slavery, and oppression of the Yanke Factorys he made proposals to come back and Manufacture some for himselfe. An agreement was made, and he bought 3 power looms – got them to the place – had the Drums put up already for belts – and I believe bought some wool, when Lo and Behold he dicovered that Jabez and his daughter Martha was grown to intimate! The project accordingly fell through, and whether it will revive again or not I can not tell. He says that he will not start the Looms except Jabez and Martha will give up Sparking. That, they will not do. I have been Informed that he further says that, Jabez and Martha may get married, that he will go to England, and that if he can not sell his looms he will BURN them. Martha is very Religious, and Jabez goes with her to meeting every Sunday, and somtimes on the week days. Religion is the order of the day – it has become as fashionable to go to meeting, as it is to wear dickeys and false Collars to hide a dirty shirt. I do not go to the meeting, because I do, and ever did, despise the fashion. I am told that it is nessacery to go to meeting for the purpose of keeping up appearances and to help us to Slip thro' the world better, but if I cannot Slip thro' the world without putting on the Smooth Cloak of Hypocricy then I will stick in it.

John had a letter from Willm Lockwood a week or 2 before I got yours. James Hirst lived here last winter, but went to Southbridge at the 1st of April. I then engaged to do the

[63]He may have gone to work for Waters and Goodell, who made broadcloths there. *History of Worcester County* 2: 106, 109.

scouring, fulling & finishing at 2¼ cents per yd. It is small wages, but they would not give any more. If I needed any help I was to find it, Save only they agree'd to help to tenter. This they did for several weeks — till at last they begun to shuffle of and said they never agree'd to do it. I think likely I shall write a verse or two on this subject and give it to them.

> For when men gets proud, and their power they abuse,
> Then what resource have I, but to fly to my muse.

We do not use any teazels, but do all our Napping with Cards, though I am perswaided some teazels would be better.[64]

I have had some thoughts of coming to see you this Summer. I should like to see you very well, and my aunt too, as I have not seen her in this Country. But I shall waite untill you write again, when I hope you will give me directions how to come. (If you should like to have me come) for I do not like the steam boat traveling, because there is so many accidents. If I could see you I could tell you more than I can write in a dozen letters. I believe we are all in good health at present and I hope you are the same. I hope you will pardon the blundering manner and shortness of this letter, because the weather is hot and my head aches a little.

Write as soon as possible. Direct as before with this addition, 'not to be opened by any other person'.

I subcribe my selfe your Nephew

Joseph Hollingworth

[To] Mr. William Rawcliff,
at Wadsworth's Factory,
Poughkeepsie,
Duches County,
New York.

---

[64]See fn. 25.

Bradley Clay
to William Rawcliff

Huddersfield 1 August, 1831

Sir:

I recieved your letter of the 17th of April on the 2nd of June, and I immediately set about finding a customer for your property, and which I have now brought to a close. I did not think it the best mode to put it to public auction. I therefore sold it by private contract, and the purchaser is Mr. Elihu Hobson of Netherthong – the price two hundred Eighty one pounds to be paid in three months, if the title prove good, of which I will advise you as soon as I know, and you then may draw upon me for the balance.

I hope you will approve of what I have done for I assure you two individuals who know the property well, said it would be well sold, if I got £ 250 for it.

The Club is not yet out, there will be two quarters yet to pay. I have had presented to me a note from George Lockwood and Isaac Brearley, amount £ 3. 1s. 6 paid by them on your account to the Netherthong old money Club. Am I to repay them?

We have had a very fine Spring, and the crops in the ground are very heavy. If the weather prove equally favourable for the harvest it is thought wheat will be down at five shillings per bushel.

All our Manufacturies are busy, yet prices do not get up. Neither will they in my opinion; it is not surprising that we undersell all this World and shall continue to do so if food continue cheap with us.

Yours respectfully
Bradley Clay

To Mr. Wm. Rawcliffe,
Poughkeepsie,
New York,
U.S.

John Hollingworth to S. A. Hitchcock [65]

Woodstock Oct 10th 1831

Mr S. A. Hitchcock

Sir

Being unwell I could not come to Southbridge to Day to attend to the business you mentioned to my Brother but if I should get Better I shall try to come in a few Days to attend to it. My Brother mentioned that you wanted some Listing making Soon and as we have been puting on new Cards and are intending to get another Carding Machine to work the Listing on we dont want to make a large Lot at present and what we do make at this time we want you to let it all be American Wool So that it will not spoil our cards.[66]

Yours with Respct
John Hollingworth

[To] Mr S. A. Hitchcock
Southbridge

---

[65]Samuel A. Hitchcock, of Tiffany, Sayles and Hitchcock, was the Agent for the Hamilton Woolen Company. "Hamilton Woolen Company. . . ."

[66]Listing is a contemporary term for a strip of cloth. Samuel Johnson, comp., *A Dictionary of the English Language* (Philadelphia, 1818). Exactly what it means in this context is unclear to me. By "putting on new cards," John Hollingworth refers to the necessary periodic renewal of the card clothing (i.e. the leather or fabric with protruding wires) that covered the carding machine cylinders. The alternative to American wool was probably imported Merino wool, which could have caused trouble in two ways. Because of its quality – it was about twice as fine as common

Joseph Hollingworth
to William Rawcliff

Pond Factory, Woodstock, Land of steady habbits
12th. month, 20th Day. 1831.

Dear Uncle

I recd. yours of the 4inst. on the 12th. I had waited a long time in expectation for it, at last I began to think that you was put out with me and would not write. The reason why, I could not account for, except that Joseph Haigh might have told you something when he was at Poughkeepsie. By this statement you may guess how pleased I was, when I got your letter, to find that I might have been entertaining wrong ideas.

When J. Haigh got home I went to see him, and found by his conversation that he had been at Poughkeepsie, but he nevver said a word to me about you, and so I didn't trouble him with any questions. So much for Friendship! When Joseph came home he found that Bro. Jabez and his daughter Martha had entered another State. Not the state of Massachusetts, nor the state of Rode Island nor any other of "these United States." But the UNITED STATE OF MATRIMONY! They was married on the 1st of Sept.bre [67]

English or American varieties — Merino wool required a closer adjustment of the carding machine. William Draper Lewis, *Our Sheep and the Tariff* (Philadelphia, 1890), especially Table 1, between pp. 60 and 61; and Cole, *Wool Manufacture* 1: 98-99. In addition, as Rees warned: "Spanish wool in the bale has generally some part of the pitch employed to mark the sheep still adhering to it, which must be carefully cut off." "Woolen Manufacture," *The Cyclopaedia*, vol. 40. These impurities in the imported wool could have clogged the card clothing.

[67]Jabez and Martha Haigh were married in Woodstock. *Vital Records.*

Jabez might think he should have plague
Were he to live a single life,
And so he did take Martha Haigh
To be his Dear and Lawful Wife.

Friend Joseph also found that in gratifying his spleenic temper he had distributed a few of his Dollars amongst the Yankees. This he did not like. So he agreed with our folks here to have his old work again, and to run his Satinet looms. He went to work as usal, but has not yet started the looms, nor do I believe that he ever intended to do. I thought of coming to see you before winter, but I waited for your letter first, and then it was too late. The winter here set in in Nov.br and has been very severe some of the time scince.
I have been requested by Father to give you an especial invitation to come over here, to see us and this place as he thinks it would be benificial to both sides. I therefore, in accordance to his wishes do ernestly invite you to come over, if you can make it convenient for yourself and family. If you should come, you might perhaps have an oppertunity of getting work in this part of the country if you prefered it. If not I would return with you, or follow after, (if it should be agreable to you) and settle down with you on a farme, if we should think it advisable. For my own part I have no doubt, but that farming might do well, if well managed. I think it is the surest (if not the easiest) way of getting a livelihood. I think that a small farme, with a mill on it, for doing custom work such as Carding and Cloth-dressing; (or a saw mill and Grist mill) would do very well for such folks as you and I. I do not exactly know for what reason our folks are so very anxious to have you come here, but I think that they want you to keep a store in New York, and to act as Agent in selling off, all, or part of the goods manufactured here. This I am constrained to believe, from something that my Wife told me, that she had heard from Sarah, (John's Wife).

GRACIOUS HEAVENS! what have I now written!! my WIFE!!! Well, I vow, I've let the cat out of the bag. So now for an explanation.

The Job is done, the Knot is tied,
And now I've got myself a bride;
In spite of some folk's worthless strife,
I've took a Yankee for my wife.

We was married on the 15th of Novbr. by the Rev. Alban Underwood of West Woodstock.[68] I am the fourth son of my Mother's that has got married,

[one page missing]

❧ ❧ ❧

Jabez Hollingworth to William Rawcliff

Sturbridge October 7th 1832

Dear Uncle

We received your Letter of the 18th of September on the 30th by which letter we learn that you had not received any letter from my Father according to his promise. The reason why he did not write sooner was because we could not get the money as soon as we expected, and when we got it, it was in

[68]Joseph married Hannah Blackemor. Rev. Alvan Underwood was pastor of the West Woodstock Congregational Church there. Woodstock, *Vital Records;* Clarence W. Bowen, *Genealogies of Woodstock Families* (Worcester, Mass., 1943), 8: 416.

such bills as was not prudent to send. Brother James had to go to Worcester Bank to get it exchanged, So that it was the 9th of September when it was sent of or put in the Post Office. On the 9th of September we sent you a letter with a Fifty dollar bill enclosed in it. Directed to William Rawcliff Rosendale Ulster County N.Y.[69] sealed with black sealing wax and stamped with a cent, written on one corner with Care and despatch, and put in as a double Letter. Now we hope that you have received that letter long before now. If you have you must write and let us know as we are very anxious to know respecting it. If you have not received it you must not loose one hour of time in enquiring at your post office and finding out all that is possible concerning it and if you can hear nothing of it you must write immediately and let us know so that we shall know what to do.

We are glad to hear that your health has improved, and we hope that your health will be continued and that you will be spared to be a blessing to your Family. Though Friends may forsake and Relations may turn against you, and Losses and disappointments may threaten to over whelm you in the Flood of despair, yet you are not alone. I myself am pretty much in the same situation in some respects. I have a Wife whom I love as dear as my own life, and whose Father and Brothers are set against us because we are poor or else because she has done the 99th good turn and has not done the 100th, who have lived here nearly 3 months and they have not come to see us excepting young Joseph and that was when he had been with his Father to Monson [Mass.] when he was going into York State and he the Father Rode within 20 yards of our door and where I was to work but would not call to see us. They have often said they would come to see us, but they cannot or will not spend time

---

[69]He was probably employed by the Rosendale [Wool] Manufacturing Company. Williams, *New York Annual Register*, p. 153.

though the distance is only about 8 miles and they have had no work of any consequence this 5 months.

Mr Haigh says that you are as big a Rascal as any of us because you have not answered the Letter that he left you and that he should not have Hired Delhi Factory[70] had it not been for you and that he should not have gone into York State if you had not invited him. He now says he will not go to Delhi. But he has given Cousin Joseph Kenyon an invitation to go with him.[71] He has offered the same offers to him which he has offered to you and if he will accept them he will go. The truth is he wants a stepping stone, a Horse to Carry the Saddle. If you are on your Land send us all particulars. I want to keep up a correspondence with you concerning your building. I should like to say more on this subject but cannot at present. My wife

---

[70]The Delaware Woollen Factory, in Delhi, Delaware County, New York. Williams, *New York Annual Register*, pp. 148, 155. For a contemporary view, see Barber and Howe, *Historical Collections of . . . New York . . .*, p. 127.

[71]Joseph Kenyon stayed in New England. Sometime after this date he leased a mill in Southbridge; he later moved to Woodstock, Connecticut and purchased a mill with 10 acres of land and the following inventory of machinery: 2 cassimere gigs, 1 shearing machine, 1 brushing machine, a gross of gig flats, 1 press, 2 carding machines, ½ gross press papers, 1 picking machine, a 180-spindle jack, 5 cassimere power looms, 1 warping machine, 1 spooling machine, a 144-spindle jack, and 100 spools. The village around the factory became known as Kenyonville. Joseph also served a term as representative to the General Assembly in Connecticut. After his death, his sons ran what was known as the Kenyon Brothers Mill. Kenyons fought on the Union side in the Civil War, and in 1920 when Woodstock produced a pageant celebrating the 300th Anniversary of the pilgrims' landing at Plymouth, two Kenyons took roles as "Colonials" in the pageant. Woodstock, *Land Records* 23: 336; Barlow and Bancroft Insurance Survey #7732 (Library, Merrimack Valley Textile Museum); Bowen, *History of Woodstock*, pp. 360, 599, 607, 622.

got her bed on the 16 September of a son. His Name is George Washington. Brother Joseph had a son about 4 weeks before. I do not know his Name. We are in pretty good health at present. I expect we shall not Finish the Meeting house before December.[72]

Be sure you attend to the letter and the Money. It has Sturbridge Post Mark on it. I am with Due Respect

Jabez Hollingworth

Direct to me at Sturbridge
Worcester County Mass.

[To] Mr William Rawcliff
High Falls Rosendale
Ulster County New York
if not there please to forward it to
Liberty
Sulivan County N.Y.

---

[72]Jabez presumably moved to Sturbridge to build this new church for the Baptists. Dedicated in January 1833, the building was moved five years later to a new site donated by Josiah I. Fiske, proprietor of the Sturbridge Cotton Mills, *History of Worcester County* 2: 366-369. For a contemporary view of this village, see John Warner Barber, *Historical Collections, . . . Of Every Town in Massachusetts, . . .* (Worcester, 1839), p. 608.

John Hollingworth to Hamilton Woolen Company

Woodstock May 2. 1837

Hamilton Woolen Co

Bot of John Hollingworth

One Single Card Machine & Condenser built by Dow & Madison		500.
One 160 Spindle Jack,	built by Kimball & Fuller	290.
One Narrow Gig & Setts Flats	" Curtis & Henshaw	80.
One " Brush Machine	" " d°	45.
One Tube Condenser attached to the 24 in. Finisher		50.
One Picker built by Chas. Midgely		60.
One Broad Loom		90.
		$1115.00

The above Machinery all now in the "Pond Factory"
so called –

John Hollingworth

Ent 2

GENEALOGICAL CHART

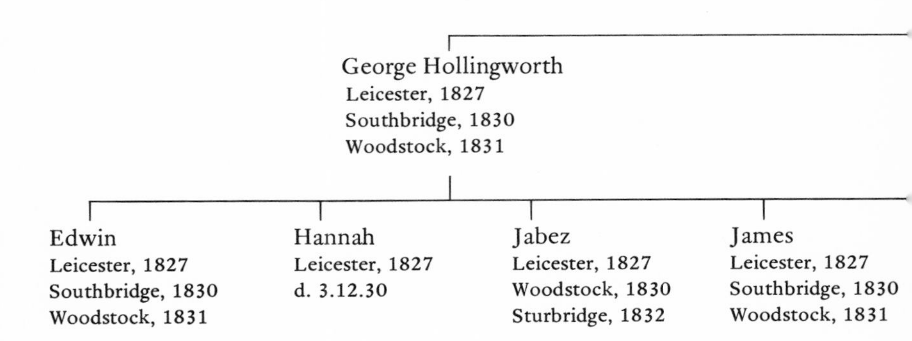

WILLIAM RAWCLIFFE (George Hollingworth's brother-in-law)

Poughkeepsie, 1829
Rosendale or Liberty, 1832

William — Mary Ann — Anis

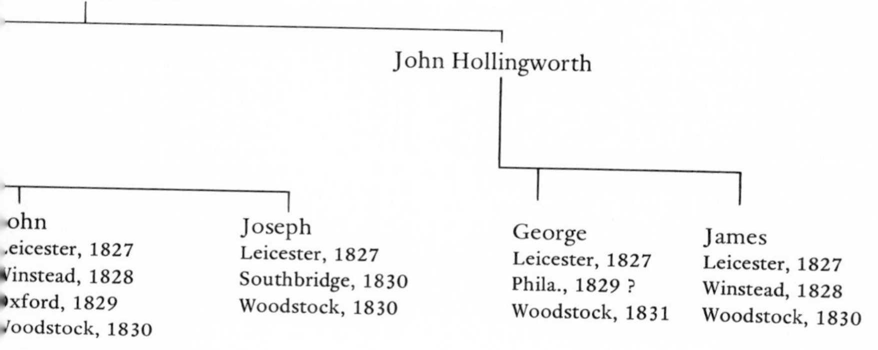
OLLINGWORTH
John Hollingworth
ohn
eicester, 1827
Vinstead, 1828
xford, 1829
Voodstock, 1830
Joseph
Leicester, 1827
Southbridge, 1830
Woodstock, 1830
George
Leicester, 1827
Phila., 1829 ?
Woodstock, 1831
James
Leicester, 1827
Winstead, 1828
Woodstock, 1830

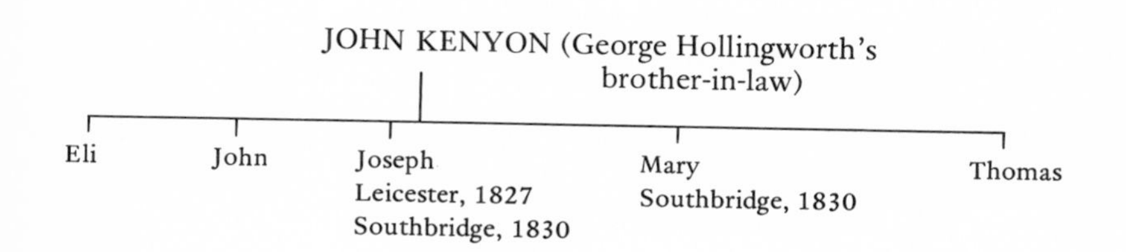
JOHN KENYON (George Hollingworth's brother-in-law)
Eli
John
Joseph
Leicester, 1827
Southbridge, 1830
Mary
Southbridge, 1830
Thomas

Bibliography

Primary Sources

Allen, Zachariah. *The Practical Tourist, or Sketches of the State of the Useful Arts, and of Society, Scenery, &c. &c. in Great-Britain, France and Holland.* 2 vols. Providence, R.I., 1832.

American Advertising Directory for Manufacturers and Dealers in American Goods. New York, 1831.

Barber, John Warner. *Connecticut Historical Collections, Containing A General Collection Of Interesting Facts, Traditions, Biographical Sketches, Anecdotes, etc., Relating To The History And Antiquities Of Every Town In Connecticut, With Geographical Descriptions.* "Improved Edition." New Haven, 1838.

———. *Historical Collections, Being A General Collection Of Interesting Facts, Traditions, Biographical Sketches, Anecdotes, &c., Relating To The History And Antiquities Of Every Town In Massachusetts, With Geographical Descriptions.* Worcester, 1839.

Barber, John W[arner], and Howe, Henry. *Historical Collections Of The State Of New York; Containing A General Collection Of The Most Interesting Facts, Traditions, Biographical Sketches, Anecdotes, &c. Relating To Its History And Antiquities, With General Descriptions of Every Township In the State.* New York, 1841.

[Dwight, Theodore, Jr.] *The Northern Traveller: (Combined with the Northern Tour.) Containing the Routes to Niagara, Quebec, and the Springs. With the Tour of New-England, and the Route to the Coal Mines of Pennsylvania.* 3d ed. rev. New York, 1828.

Ellis, Asa. *The Country Dyer's Assistant.* Brookfield, Mass., 1798.

Evans, Oliver. *The Young Mill-Wright and Miller's Guide.* 6th ed. Philadelphia, 1829.

Hall, Basil. *Travels in North America in the Years 1827 and 1828.* 3 vols. Edinburgh, 1829.

[Hamilton, Thomas.] *Men and Manners in America.* Philadelphia, 1833.

Hamilton Woolen Company, Southbridge, Mass. *Records.* Baker Library, Harvard University.

Hazen, Edward. *The Panorama of Professions and Trades.* Philadelphia, 1837.

Journal of the Franklin Institute. Vols. 1-13 (new series). Philadelphia, 1828-1834.

Lincoln, William. *History Of Worcester, Massachusetts, From Its Earliest Settlement To September, 1836; With Various Notices Relating To The History Of Worcester County.* Worcester, 1837.

Mechanics' Magazine and Register of Inventions and Improvements. Vols. 1-5. New York, 1833-1835.

Niles' Weekly Register. Vols. 30-43. Baltimore, 1826-1832.

O'Rielly, Henry. *Sketches of Rochester.* Rochester, 1838.

Pennsylvania, House of Representatives. *Report* of the Committee on Inland Navigation and Internal Improvement, relative to the further extension of the Pennsylvania canal, accompanied with a bill. (Read, H.J.R., February 28, 1827). 18 pp. Harrisburg, 1827.

Scrope, G. Poulett. *Extracts of Letters From Poor Persons Who Emigrated Last Year to Canada and the United States.* 2d ed. London, 1832.

Smith, J. Calvin. *The Western Tourist and Emigrant's Guide.* New York, 1840.

Southbridge, Mass. *Vital Records.*

Spofford, Jeremiah. *A Gazetteer of Massachusetts.* Newburyport, 1828.

Stuart, James. *Three years in North America.* 3d ed. rev. 2 vols. Edinburgh, 1833.

Tanner, H. S. *A Brief Description of the Canals and Railroads of the United States.* Philadelphia, 1834.

Tocqueville, Alexis de. *Democracy in America.* Translated by Henry Reeve, preface and notes by John C. Spencer. 4th ed. 2 vols. New York, 1841.

U.S., Congress, House, Committee on Manufactures. *Report on Petition relative to . . . Duties on Imports.* 20th Cong., 1st sess., January 31, 1828, H. Rept. 115, vol. 2.

U.S., State Department, "Questions to be addressed to the Persons concerned in Manufacturing Establishments. . . ." manuscript returns, Fourth Census (1820), National Archives. Selected Xerox copies in Merrimack Valley Textile Museum.

U.S., Treasury Department, *Documents Relative to the Manufactures in the United States.* 22d Cong., 1st sess., 1833, House Executive Document 308. 2 vols.

Ure, Andrew. *A Dictionary of Arts, Manufactures, and Mines; Containing a Clear Exposition Of Their Principles and Practice.* 2 vols. New York, 1845.

Wansey, Henry. *The Journal of an Excursion to the United States of North America in the Summer of 1794.* Salisbury, England, 1796.
White, George S. *Memoir of Samuel Slater.* 2nd ed. Philadelphia, 1836.
Williams, Edwin. *The New York Annual Register for the Year of Our Lord 1830. . . .* New York, 1830.
Woodstock, Connecticut. *Land Records.*
______. *Vital Records.*
"Wool" and "Woollen Manufacture," *The Cyclopaedia; or Universal Dictionary of Arts, Sciences, and Literature,* edited by Abraham Rees. Vol. 40. Philadelphia, 1822.
"Wool" and "Woollens," *Encyclopedia Americana,* edited by Francis Lieber. 13: 248-252, 252-254. Philadelphia, 1830-1833.

Secondary Sources

Ammidown, Holmes. *Historical Collections.* 2 vols. New York, 1874.
Aspin, C., and Chapman, S. D. *James Hargreaves and the Spinning Jenny.* Helmshore, Lancashire: Helmshore Local History Society, 1964.
Bagnall, William R. *Sketches of Manufacturing Establishments in New York City, and of Textile Establishments in the Eastern States.* Edited by Victor S. Clark. 4 vols. Unpublished manuscript, 1908. Contributions to American Economic History. Washington, D. C.: Carnegie Institution of Washington. Xerox copy in Merrimack Valley Textile Museum.
______. *The Textile Industries of the United States.* Cambridge, Mass., 1893.
Bayles, Richard M., ed. *History of Windham County.* New York, 1889.
Bishop, J. Leander. *A History of American Manufactures from 1608 to 1860.* Vol. 2. Philadelphia, 1864.
Bowen, Clarence Winthrop. *The History of Woodstock, Connecticut.* Norwood, Mass.: Privately printed, 1926.
"Brief Record of the History of the Hamilton Woolen Company, Southbridge, Massachusetts." Typescript (copy) filed with the company records, Baker Library, Harvard University. 3 pp.
Calvert, Monte A. "The Technology of the Woolen Cloth Finishing Industries from Ancient Times to the Present, With Special Emphasis on American Developments 1790-1840, and on the Processes of Fulling, Napping and Shearing." Unpublished manuscript. 89 pp. Xerox copy in Merrimack Valley Textile Museum. North Andover, 1963.

Cole, Arthur H. *The American Wool Manufacture.* 2 vols. Cambridge, Mass.: Harvard University Press, 1926.

Cowley, Charles. *Illustrated History of Lowell.* Rev. ed. Boston, 1868.

Craik, David. *The Practical American Millwright and Miller.* Philadelphia, 1870.

Crump, W. B., and Ghorbal, Gertrude. *History of the Huddersfield Woollen Industry.* Handbook 9, The Tolson Memorial Museum Publications. Edited by T. W. Woodhead. Huddersfield, 1935.

Daniels, George F. *History of the Town of Oxford, Massachusetts.* Oxford, 1892.

Habakkuk, H. J. *American and British Technology in the Nineteenth Century.* New York: Cambridge University Press, 1962.

Hartwell, R. M. "The Yorkshire Woollen and Worsted Industries, 1800-1850." Unpublished D. Phil. thesis, Oxford, 1956.

History of Worcester County, Massachusetts. 2 vols. Boston, 1879.

Hotten, John C. *Original Lists of Emigrants, 1600-1700.* London, 1874.

Kirkland, Edward C. *A History of American Economic Life.* Crofts American History Series. General Editor, Dixon Ryan Fox. New York: Appleton-Century-Crofts, Inc., 1939.

Larned, Ellen. *History of Windham County, Connecticut.* 2 vols. Worcester, 1874, 1880.

Meyer, Balthasar H. *History of Transportation in the United States Before 1860.* Contributions to American Economic History. Washington, D. C.: Carnegie Institution of Washington, 1948.

North, S. N. D. "The New England Wool Manufacture," *Bulletin of the National Association of Wool Manufacturers* 29 (1899): 113-156, 214-268; 30 (1900): 23-65, 119-163, 213-263, 317-349; 31 (1901): 168-208, 253-286, 390-416; 32 (1902): 101-146, 305-334; 33 (1903): 143-178.

Pursell, Carroll W., Jr. "Thomas Digges and William Pearce: An Example of the Transit of Technology." *William and Mary Quarterly,* 3d ser. 21, no. 4 (October 1964): 551-560.

Quinabaug Historical Society Leaflets. 3 vols. Southbridge, Mass.: n.d.

Quinlan, James Eldridge. *History of Sullivan County.* Liberty, N.Y., 1873.

Redford, Arthur W. *Labour Migration in England, 1800-1850.* 2d ed., edited by W. H. Chaloner. Manchester: Manchester University Press, 1964.

Riznik, Barnes. "New England Wool-Carding and Finishing Mills,

1790-1840." Unpublished manuscript. 125 pp. Carbon copy in Merrimack Valley Textile Museum. Sturbridge, 1964.

Rogers, Grace L. "The Scholfield Wool-Carding Machines." In *Contributions from the Museum of History and Technology*, pp. 1-14. United States National Museum Bulletin 218. Washington, D. C.: Smithsonian Institution, 1959.

Rosenbloom, Richard S. "The Technology of Woolen Spinning. Its History and Development; With Emphasis on the American Woolen Industry in the Nineteenth Century." Unpublished manuscript. 49 pp. Mimeograph copy in Merrimack Valley Textile Museum. North Andover, 1962.

Shepperson, Wilbur S. *British Emigration to North America.* Minneapolis: University of Minnesota Press, 1957.

———. *Emigration and Disenchantment: Portraits of Englishmen Repatriated from the United States.* Norman: University of Oklahoma Press, 1965.

Smith, James H. *History of Dutchess County, New York.* Syracuse, 1882.

Temple, J. H. *History of Framingham, Massachusetts.* Framingham, 1887.

U.S. Bureau of the Census. *Historical Statistics of the United States. Colonial Times to 1957.* Washington, D. C., 1960.

Index

Agrarian impulse, evidence of, 12, 14, 24, 25, 45, 51, 58, 82, 93, 98-99, 104
America, attitude toward, 19, 20, 25, 30, 31, 37, 38
Antimasonry, 76, 93

Canals, 6, 10, 16, 51
Carding mills, 85-86 n
Cloth-making equipment, xviii-xix, xxiv, 48 n, 49, 67, 68, 74 n, 76, 80, 88-89, 107 n, 109
Cloth-making processes, xviii n
  carding, 7 n
  condensing, 27 n
  napping, xix n, 27 n, 70, 100
  pressing, 67 n
  roping, 77 n
  shearing, 40 n
  slubbing, 7 n
  spinning, 13, 26 n-27 n
  tentering, 22 n
  warp winding, 28 n
  weaving, 13 n, 43 n, 50 n, 69
  weft winding, 27 n
Coal, 35, 51

England
  attitude toward, 19, 20, 31, 38
  economic conditions in, xix, 90, 101
  emigration from, xvi-xvii

Factory system, xviii-xxii
  attitude toward, 8, 9, 15, 23, 28, 34, 38, 57, 64, 66, 99
  housing for, 26
  labor practices in, 22, 26 n, 34, 40, 42, 53, 55, 58, 61, 66
  living conditions, 15
  wages, 15, 17, 26, 27, 29, 39, 61, 66, 88, 93, 100

Goulding, John, 27 n

Hudson River, 11

Lake Erie, 10-11

Machinery manufacturers, xx n, 37 n, 70, 109
Merino wool, 102 n-103 n

New York (state), 6, 11, 14, 16

Owen, Robert, xxii, 5
  influence of, 7, 16

Prices, 66
  of clothing, 16, 39
  of farms and farmland, 45
  of food, 3-4, 15-16, 34, 79, 83
  of machinery, 70, 109
  of mills, 67, 68, 80, 81, 89, 95
  of transportation, 3-4, 56

Religion
  beliefs, 62, 63
  practices, 34, 53-54, 72, 98, 99, 105
Report of the Commission on Inland Navigation and Internal Improvement (Pennsylvania House of Representatives), 10
Rochester, N.Y., 11

Steam boats, 10-11, 100

Taxes, 15, 32, 34
Temperance, 30, 32, 79
Textile mills
  in Delhi, N.Y., 107
  in Framingham, Mass., 13n
  in Herkimer County, N.Y., 37
  in Leicester, Mass., xxiii-xxiv
  in Millbury, Mass., 99
Textile mills (*continued*)
  in Oxford, Mass., 49n, 66n
  in Pine Grove, N.Y., 34
  in Pleasant Valley, N.Y., 11
  in Poughkeepsie, N.Y., 19n
  in Southbridge, Mass., xxiv, 74n
  in Sturbridge, Mass., 108n
  in Walden, N.Y., xxn
  in Winchester, Conn., 37n
  in Woodstock, Conn., xxiv-xxv, 74, 97, 109
  in Worcester, Mass., 70
Travel conditions, 21, 35
Turning mills, 81n

Utica, N.Y., 16

Water wheel, over-shot, 89
Watts, Isaac, 8